ENERGY POLICIES, POLITICS AND PRICES

THE WEATHERIZATION ASSISTANCE PROGRAM

ENERGY POLICIES, POLITICS AND PRICES

Additional books in this series can be found on Nova's website under the Series tab.

Additional E-books in this series can be found on Nova's website under the E-book tab.

LAWS AND PROGRAMS

Additional books in this series can be found on Nova's website under the Series tab.

Additional E-books in this series can be found on Nova's website under the E-book tab.

ENERGY POLICIES, POLITICS AND PRICES

THE WEATHERIZATION ASSISTANCE PROGRAM

SHERWYN L. KING
AND
DALLY R. GRIFFITHS
EDITORS

Nova Science Publishers, Inc.
New York

For permission to use material from this book please contact us:
Telephone 631-231-7269; Fax 631-231-8175
Web Site: http://www.novapublishers.com

Additional color graphics may be available in the e-book version of this book.

Library of Congress Cataloging-in-Publication Data

ISBN: 978-1-62081-745-2

Published by Nova Science Publishers, Inc. † New York

CONTENTS

PREFACE

The DOE Weatherization Assistance Program enables low-income families to permanently reduce their energy bills by making their households more energy efficient. The DOE program guidelines specify that a variety of energy efficiency measures are eligible for support under the program. The measures include insulation, space-heating equipment, energy-efficient windows, water heaters, and efficient air conditioners. This book explores the Department of Energy's (DOE's) Weatherization Assistance Program, with a focus on its background, brief history of funding, program evolution, program activity, and program assessments and benefit-cost evaluations.

Chapter 1 - This report analyzes the Department of Energy's (DOE's) Weatherization Assistance Program. (WAP, the "program"). It provides background—a brief history of funding, program evolution, and program activity—and a review of program assessments and benefit-cost evaluations.

Budget debate over the program is focused on a $5 billion appropriation in the Recovery Act of 2009, a report that state and local governments have yet to commit about $1.5 billion of that total, and concerns about the quality of weatherization projects implemented with Recovery Act funding. During the debate over FY2011 funding, the House Republican Study Committee called for program funding to be eliminated. In April 2011, Congress approved $171 million for the program in the final continuing resolution for FY2011 (P.L. 112-10). For FY2012, DOE requested $320 million, the House approved $30 million, the Senate Appropriations Committee recommended $171 million, and the Conference Committee approved $68 million. The budget debate provides the context for this report, but details of the current debate are beyond the scope of this report, which is focused on evaluations of program cost-effectiveness.

WAP is a formula grant program: funding flows from DOE to state governments and then to local governments and weatherization agencies. Over the 32 years from the program's start-up in FY1977 through FY2008, Congress appropriated about $8.7 billion (in constant FY2010 dollars). The $5 billion provided by the Recovery Act added more than 50% to the previous spending total.

Over the program's history, DOE's Oak Ridge National Laboratory (ORNL) and the Office of Management and Budget have used process and impact evaluation research methods to assess WAP operations and estimate cost-effectiveness. Virtually all the studies conducted through 2005 showed that the program was moderately cost effective. The studies included measures of operational effectiveness, energy savings, and non-energy benefits. The timing of past studies was a bit sporadic, driven mainly by new statutory requirements and program audits. Performance assessments have alternately identified improvements in program operations or identified operational problems that subsequently stimulated program improvements. Only the intensive evaluation study of program year 1989 (published in 1993) was designed to directly draw a national sample to produce empirical data on program cost-effectiveness. Such in-depth evaluations are costly and time-consuming. Most other "metaevaluations" were much less costly, using available state-level evaluation studies as the basis to infer national-level program impacts.

The large infusion of Recovery Act funding—and attendant changes in program structure— heightened interest in conducting a fresh assessment of operations and new scientifically based evaluations of program impacts. Unforeseen recession-driven events delayed use of Recovery Act funding. For example, the DOE Inspector General (DOE IG) found that recession-driven budget shortfalls, state hiring freezes, and state-wide planned furloughs delayed program implementation—and created barriers to meeting spending and home weatherization targets. In December 2011, the Government Accountability Office (GAO) released a performance audit which found that the Recovery Act phase of the program was successfully addressing most goals and the challenges identified by the DOE IG. Recently-launched evaluation studies by DOE aim to determine whether Recovery Act funding was used cost-effectively and whether it fulfilled goals for job creation.

The report concludes by reviewing the debate over the use of "outside" contractors to improve the objectivity and independence of weatherization program evaluations.

Chapter 2 - The American Recovery and Reinvestment Act of 2009 (Recovery Act) provided $5 billion to the Department of Energy's (DOE) Weatherization Assistance Program to help low-income families by making long-term energy efficiency improvements to their homes. The Recovery Act requires GAO to conduct bimonthly reviews of how recipients such as state-level agencies use the act's funds. As part of this review, GAO examined if the act is achieving its stated purposes. The act also requires GAO to comment and report quarterly on estimates of jobs funded and counted as full-time equivalents (FTE), as reported by recipients of Recovery Act funds. GAO examined (1) the status and use of weatherization grant program funds under the Recovery Act; (2) the challenges, if any, that recipients faced in implementing the weatherization program under the Recovery Act; (3) the extent to which the weatherization program under the Recovery Act has achieved its energy and cost savings goals; and (4) the changes, if any, over time in the quality of FTE data reported by Recovery Act recipients, particularly by program recipients. GAO surveyed the 58 state-level grant recipients of the act's weatherization funds, reviewed DOE and recipient-reported data, and interviewed state and local agency officials.

GAO makes no new recommendations in this report but provides the status of prior recommendations that remain open and not implemented from GAO's Recovery Act–mandated reports. DOE generally concurred with GAO's findings and provided clarifications, which were incorporated as appropriate.

Chapter 3 - This technical memorandum is intended to provide readers with information that may be useful in understanding the purposes, performance, and outcomes of the Department of Energy's (DOE's) Weatherization Assistance Program (Weatherization). Weatherization has been in operation for over thirty years and is the nation's largest single residential energy efficiency program. Its primary purpose, established by law, is "...to increase the energy efficiency of dwellings owned or occupied by low-income persons, reduce their total residential energy expenditures, and improve their health and safety, especially low-income persons who are particularly vulnerable such as the elderly, the handicapped, and children."

The American Reinvestment and Recovery Act PL111-5 (ARRA), passed and signed into law in February 2009, committed $5 Billion over two years to an expanded Weatherization Assistance Program. This has created substantial interest in the program, the population it serves, the energy and cost savings it produces, and its cost-effectiveness. This memorandum is intended to address the need for this kind of information.

Statistically valid answers to many of the questions surrounding Weatherization and its performance require comprehensive evaluation of the program. DOE is undertaking precisely this kind of independent evaluation in order to ascertain program effectiveness and to improve its performance. Results of this evaluation effort will begin to emerge in late 2010 and 2011, but they require substantial time and effort.

In the meantime, the data and statistics in this memorandum can provide reasonable and transparent estimates of key program characteristics. The memorandum is laid out in three sections. The first deals with some key characteristics describing low-income energy consumption and expenditures. The second section provides estimates of energy savings and energy bill reductions that the program can reasonably be presumed to be producing. The third section deals with estimates of program cost-effectiveness and societal impacts such as carbon reduction and reduced national energy consumption.

Each of the sections is brief, containing statistics, explanatory graphics and tables as appropriate, and short explanations of the statistics in order to place them in context for the reader. The companion appendices at the back of the memorandum explain the methods and sources used in developing the statistics.

Chapter 4 - The American Recovery and Reinvestment Act of 2009 (Recovery Act) was enacted to stimulate the U.S. economy, create jobs and make infrastructure investments in energy and other areas. The Department of Energy's (Department) Weatherization Assistance Program received about $5 billion under the Recovery Act to improve the energy efficiency of nearly 590,000 residences of low income citizens -- a dramatic increase over the $450 million appropriated for this Program in Fiscal Year 2009. The Department awarded $4.73 billion of the Recovery Act funding in the form of grants to all 50 states, 5 territories, the District of Columbia and 2 Native American Tribes (the remaining $270 million was retained by the Department to administer the grants). It was anticipated that the expenditure of these funds would result in the almost immediate creation of jobs because they were to rely on the existing Weatherization Program infrastructure, recognizing that many homes of low income citizens were in need of energy usage improvements. As of December 2009, grantees had been authorized to spend up to 50 percent of awarded funds, with additional monies to be provided based on performance. Corresponding to funding increases, weatherization production goals increased significantly, rising from 104,000 in 2009 to 586,015 units over the 3 year life of the Recovery Act.

As noted in our *Special Report on the American Recovery and Reinvestment Act at the Department of Energy* (OAS-RA-09-01, March 2009), the Department of Energy's grant-making authority, an essential component of the Weatherization Program's strategy, was critical to achieving the desired Recovery Act-related economic stimulus. In that report, we suggested that the Department take steps to develop aggressive safeguards to ensure that performance was monitored throughout the life-cycle of the grants. Because of the importance of this program to stimulating the economy, creating jobs, and improving the quality of life in low income households, we initiated this review to provide the Department with an interim status report highlighting factors impacting progress in meeting Weatherization Assistance Program and Recovery Act goals.

In: The Weatherization Assistance Program ISBN: 978-1-62081-745-2
Editors: S. L. King and D. R. Griffiths

Chapter 1

DOE Weatherization Program: A Review of Funding, Performance, and Cost-Effectiveness Studies*

Fred Sissine

Summary

This report analyzes the Department of Energy's (DOE's) Weatherization Assistance Program. (WAP, the "program"). It provides background—a brief history of funding, program evolution, and program activity—and a review of program assessments and benefit-cost evaluations. Budget debate over the program is focused on a $5 billion appropriation in the Recovery Act of 2009, a report that state and local governments have yet to commit about $1.5 billion of that total, and concerns about the quality of weatherization projects implemented with Recovery Act funding. During the debate over FY2011 funding, the House Republican Study Committee called for program funding to be eliminated. In April 2011, Congress approved $171 million for the program in the final continuing resolution for FY2011 (P.L. 112-10). For FY2012, DOE requested $320 million, the House approved $30 million, the Senate Appropriations Committee recommended $171 million, and the Conference Committee approved $68 million. The budget debate provides the

* This is an edited, reformatted and augmented version of the Congressional Research Service Publication, CRS Report for Congress R42147, dated January 11, 2012.

context for this report, but details of the current debate are beyond the scope of this report, which is focused on evaluations of program cost-effectiveness. WAP is a formula grant program: funding flows from DOE to state governments and then to local governments and weatherization agencies. Over the 32 years from the program's start-up in FY1977 through FY2008, Congress appropriated about $8.7 billion (in constant FY2010 dollars). The $5 billion provided by the Recovery Act added more than 50% to the previous spending total. Over the program's history, DOE's Oak Ridge National Laboratory (ORNL) and the Office of Management and Budget have used process and impact evaluation research methods to assess WAP operations and estimate cost-effectiveness. Virtually all the studies conducted through 2005 showed that the program was moderately cost effective. The studies included measures of operational effectiveness, energy savings, and non-energy benefits. The timing of past studies was a bit sporadic, driven mainly by new statutory requirements and program audits. Performance assessments have alternately identified improvements in program operations or identified operational problems that subsequently stimulated program improvements. Only the intensive evaluation study of program year 1989 (published in 1993) was designed to directly draw a national sample to produce empirical data on program cost-effectiveness. Such in-depth evaluations are costly and time-consuming. Most other "metaevaluations" were much less costly, using available state-level evaluation studies as the basis to infer national-level program impacts. The large infusion of Recovery Act funding—and attendant changes in program structure— heightened interest in conducting a fresh assessment of operations and new scientifically based evaluations of program impacts. Unforeseen recession-driven events delayed use of Recovery Act funding. For example, the DOE Inspector General (DOE IG) found that recession-driven budget shortfalls, state hiring freezes, and state-wide planned furloughs delayed program implementation—and created barriers to meeting spending and home weatherization targets. In December 2011, the Government Accountability Office (GAO) released a performance audit which found that the Recovery Act phase of the program was successfully addressing most goals and the challenges identified by the DOE IG. Recently-launched evaluation studies by DOE aim to determine whether Recovery Act funding was used cost-effectively and whether it fulfilled goals for job creation. The report concludes by reviewing the debate over the use of "outside" contractors to improve the objectivity and independence of weatherization program evaluations.

BACKGROUND

The DOE Weatherization Assistance Program enables low-income families to permanently reduce their energy bills by making their households more energy efficient. DOE program guidelines specify that a variety of energy efficiency measures are eligible for support under the program. The measures include insulation, space-heating equipment, energy-efficient windows, water heaters, and efficient air conditioners.

Statutory Authority

The program was created under Title IV of the Energy Conservation and Production Act of 1976 (P.L. 94-385).[1] The statute specifies that the program's primary purpose is

> to increase the energy efficiency of dwellings owned or occupied by low-income persons, reduce their total residential energy expenditures, and improve their health and safety, especially low-income persons who are particularly vulnerable such as the elderly, the handicapped, and children.[2]

The 1973 oil crisis caused rapid increases in energy prices, which caused major economic dislocations for the nation. As one result, the program was designed to save imported oil and cut heating bills for low-income households. This included senior citizens living on fixed incomes and Social Security, who were especially hard hit by rising energy bills.[3]

The Department of Health and Human Services (HHS) operates a Low-Income Home Energy Assistance Program (LIHEAP), which was designed to help pay energy bills for low-income households.[4] Since the inception of LIHEAP in 1981, up to 15% of funds could be used for weatherization. In 1990, the statute was amended to allow states to use up to 25% of funds for weatherization, without requiring a waiver from HHS.[5]

Program Structure

Types of Weatherization Services

At first, weatherization providers emphasized low-cost measures, such as covering windows with plastic sheets and caulking and weatherstripping windows and doors. Many of these activities involved emergency and temporary measures. With the accumulation of experience over time, the emphasis shifted to more permanent and more cost-effective measures. The range of qualified measures expanded to include storm windows and doors, attic insulation, space heating and water heating systems, furnace and boiler replacements, and cooling efficiency measures.[6] The cooling efficiency measures include activities such as air conditioner replacements, ventilation equipment, and screening and shading devices. DOE recounts that the use of home energy audits was key to adapting portfolio of measures:

> In the 1990s, the trend toward more cost-effective measures continued with the development and widespread adoption of advanced home energy audits. This proved to be a key advance for weatherization service providers since it required every home to be comprehensively analyzed before work began in order to select the most cost-effective measures and the best approach. This custom analysis of every home has become the hallmark of weatherization and ensures each client receives the most cost-effective treatment.[7]

Eligible Groups

DOE's Weatherization Program is one of the largest energy efficiency programs in the nation. It is implemented in all 50 states, in the District of Columbia, in the U.S. trust territories, and by Native American Tribes. Vulnerable groups are targeted, including the elderly, people with disabilities, and families with children.

A high priority is given to households with an elderly or disabled member. In FY2000, 49% of the weatherized households were occupied by an elderly resident or by a person with a disability.

Low-Income Population and Energy Cost Burden

The American Recovery and Reinvestment Act of 2009 (Recovery Act, P.L. 111-5; §407a) revised the program guidelines to raise the low-income eligibility ceiling from 150% to 200% of the poverty level.[8] Low-income households have lower total energy use and smaller bills than the non-low-income population. However, those bills represent a higher proportion of total income. For 2009, Oak Ridge National Laboratory (ORNL) estimated the average energy cost burden at about 10% of income for low-income households compared to about 3.3% for non-low-income households.[9] DOE elaborates further:

> Low-income households have lower average residential energy usage and lower residential energy bills than the non-low-income population, but this difference is not in proportion to household income. The average income of low-income households as provided in the 2005 RECS and adjusted for inflation was estimated at $18,624 compared to $71,144 for non-low-income households. In 2009 the group energy burden of low-income households, defined as average residential energy expense divided by average income, was estimated to be 10 percent of income for low-income households compared to 3.3 percent for non-low-income households.... Households that actually received energy payment assistance, estimated at just over 5 million in 2005, had an even higher energy burden of11.5 percent of income.[10]

Allocation Formula and Intergovernmental Administration

DOE employs a formula to allocate funding to the states and territories.[11] Each state and territory, in turn, decides how to allocate its share of the funding to local governments and jurisdictions.[12] Funds made available to the states are allocated to local governments and nonprofit agencies for purchasing and installing energy efficiency materials, such as insulation, and for making energy-related repairs.[13]

Funding Use Breakdown

The law directs DOE to reserve funds for national training and technical assistance (T&TA) activities that benefit all states and Native American Tribes. DOE allocates funding for T&TA activities at both the state and local levels. The total funding for national, state, and local T&TA was originally limited to 10% of an annual congressional appropriation.[14] The Recovery Act allowed the T&TA share to increase temporarily to 20%.

The remaining funds comprise the total allocation to state programs. The program allocation consists of two parts: the base allocation and the formula allocation. The base allocation for each state is fixed, but the amount differs for each state. The fixed base was computed so that a revised formula would not cause large swings from previous allocations, which could disrupt a state's program operations.[15] **Appendix A** provides a history of total annual funding.

In FY2010, a new program account for "Innovations in Weatherization" was funded. The new activity was designed to demonstrate new ways to increase the number of low-income homes weatherized and lower the federal cost per home for residential retrofits, while also establishing a stable funding base. Partnerships with traditional weatherization providers such as non-profits, unions, and contractors is the focus. The partners are expected to leverage financial resources, with a goal of $3 of non-federal contributions for each $1 that DOE provides. [16]

Distribution Factors

The distribution of total formula allocations across the states is based on three factors: the relative size of the low-income population, climatic conditions, and residential energy expenditures.

Table 1. Funding Allocation Priorities and Proportional Funding Cuts

Priority	**Program Component**	**Share of Annual Appropriation**	**After Total TTA, Remaining Funds greater than $209.7 million**	**After Total TTA, Remaining Funds less than $209.7 million**
1	TTA—National	—	—	—
2	TTA—State & Local	—	—	—
	Total TTA	Maximum 20%	—	—
3	Base Allocation	Minimum 80%	yes	yes, reduced proportionally
4	Formula Allocation		yes, based on population, climate, and energy use	no

Source: DOE, *Overview of the Allocation Formula for the Weatherization Program*, http://www.waptac.org/data/files/Website.../Allocation%20Funding.docx; "Weatherization Assistance Program for Low-Income Persons," *Federal Register*, March 25, 2009, http://www.federalregister.gov/articles/2009/03/25/E9-6628/weatherization-assistance-program-for-low-income-persons.

Notes: For further details, see "Weatherization Assistance Program for Low-Income Persons," *Federal Register*, March 25, 2009, http://www.federalregister.gov/articles/2009/03/25/E9-6628/weatherization-assistance-program-for-low-income-persons.

The low-income population factor is the share of the nation's low-income households in each state expressed as a percentage of all U.S. low-income households. The climatic conditions factor is obtained from the heating and cooling degrees for each state, treating the energy needed for heating and cooling proportionately. The residential energy expenditure factor is an approximation of the financial cost burden that energy use places on low-income households.[17]

Funding Allocation Priorities and Proportional Funding Cuts

In the event of funding cuts below a minimum threshold level, DOE program rules specify how cuts would be carried out.[18] The rule provides funding according to four priority levels. In descending order, the priorities are: national training and technical assistance (TTA) activities, TTA for state and local levels, base allocation to states, and—if the funds remaining after TTA exceed a threshold of $209.7 million—then a formula is used to spread the remainder among the states. **Table 1** illustrates these priorities. In the other case—if the funds remaining after the total TTA allocation are less than $209.7 million—then there is no formula allocation and the base allocation is reduced proportionally.[19]

FUNDING HISTORY: SELECTED HIGHLIGHTS[20]

Funding continuity has been elusive for this program. Over its history, WAP program funding has followed an up-and-down pattern, framed by occasional funding spikes and proposals to eliminate program funding and operations.

Funding Trend Highlights

The chart in **Figure B-1** of Appendix B, and the data in **Table A-1** of Appendix A, show the variation in the program's historical funding trend, calibrated in constant 2010 dollars.[21] **Table A-1** shows that the accumulated congressional appropriations for the 32-year period from FY1977 through FY2008 reached a sum of nearly $8.7 billion.[22] For FY2009, the Recovery Act made a special one-time appropriation of $5.0 billion, which added about 57% to the sum of all previous appropriations through the end of FY2008.

Figure B-1 shows the alternating pattern of support, characterized by occasional agreements— and periods of marked differences—between administration and congressional funding viewpoints. A few observations on this history:

- **Funding Range.** Except for the FY2009 Recovery Act, single-year program funding has ranged between a high of about $500 million in FY1979 to a low of about $150 million in FY1996 (constant FY2010 dollars);
- **Funding Trend.** Except for the FY2009 Recovery Act appropriation, program funding has been on a long-term downtrend since FY1979 (constant FY2010 dollars);
- **Past Requests for Termination.** In 8 out of 34 years (from FY1978 throughFY2011), the administration requested zero funding (program termination):FY1982, FY1983, FY1984, FY1987-FY1990, and FY2009;
- **Administration-Congress Agreements.** In 7 years of the 35-year funding history from FY1977 through FY2011, Congress approved nearly the same funding (less than 6% difference) level as the administration requested: FY1979, FY1980, FY1981, FY1985, FY2001, FY2006, and FY2010;
- **Request-Appropriation Percent Differences.** Including the 8 years of zero funding requests, in 20 out of 35 years, the final congressional appropriation differed (plus or minus) from the administration request by more than 20%;
- **Request-Appropriation Differences Larger than $50 Million.** In 20 out of 35 years (since FY1977), the final congressional appropriation differed (plus or minus) from the administration request by more than $50 million;
- **Appropriation Exceeded Request by $200 Million.** In 10 out of 35 years (since FY1977), Congress appropriated a funding level that exceeded the administration request by more than $200 million (in 2010 dollars): FY1982-FY1984, FY1987– FY1992, and FY2009.

Funding History Highlights, FY1977 through FY2008

Presidents Ford, Carter, and Reagan Administrations

Congress started the Weatherization Program in FY1977, during the Ford Administration. Prodded by the second oil import crisis, funding was driven rapidly upward during the Carter Administration through FY1980. For FY1981, the Carter Administration requested $428.2 million (in constant 2010 dollars), which the 96th Congress approved. In January 1981, the outgoing Carter Administration issued a DOE FY1982 budget request that sought

$402.8 million (in constant 2010 dollars) for the Weatherization Program. Shortly after coming into office in January 1981, the incoming Reagan Administration reversed the previous trend by requesting that $26.2 million be rescinded from the FY1981 appropriation. The 97th Congress approved the rescission, bringing program funding down to $376.6.0 million (in constant 2010 dollars). In March 1981, the Reagan Administration issued an FY1982 request for zero funding, noting that it planned to restructure WAP as a block grant program under the Department of Housing and Urban Development. In DOE's FY1983 request, the Reagan Administration again proposed to terminate the Weatherization Program and other DOE grant programs:

> The Energy Conservation Grants account consolidates financial and technical assistance programs carried out under the Energy Conservation appropriation which are proposed for termination in FY1983 in support of the Administration's proposal to dismantle DOE. These programs include State and Local Assistance.... The budget reductions are in response to the fact that motivated by rising energy cost and federal tax policies, individuals, businesses and other institutions have undertaken major conservation efforts. The President's economic recovery program, in conjunction with oil price decontrol and increasing natural gas prices will accelerate this trend. Public awareness of energy conservation benefits and the high level of private investment in energy conservation clearly show that the State/Local grant programs do not warrant further federal support.[23]

Despite the zero request, Congress approved nearly $473 million in 2010 dollars for FY1983. The Reagan Administration sought zero funding again in FY1984, FY1987, FY1988, and FY1989. Following FY1983, appropriations trailed downward steadily—and stood at about $258 million in 2010 dollars in FY1989.

Presidents George H. W. Bush and Clinton Administrations

In parallel to the Reagan Administration requests for zero funding in FY1987, FY1988, and FY1989, the George H.W. Bush Administration sought zero funding for FY1990 and only relatively small amounts for FY1991 ($22 million, in 2010 dollars) and FY1992 ($35 million, in 2010 dollars). In its first request, the Clinton Administration sought a major increase for FY1994 ($332 million, in 2010 dollars), and sustained the request at about that level for FY1995. Congress approved most of those two requests. For FY1996, the Administration came back with a request at nearly the same level (about $306 million, in 2010 dollars), but Congress approved only about $150 million (in

2010 dollars)—less than half the request. In real dollar terms, this was the lowest appropriation since FY1977. For FY1997 through FY2001, the Administration's requests hovered in a range from about $190 million to $200 million, in 2010 dollars. Congress generally approved $20 million to $40 million less than those requests, in 2010 dollars. For FY2001, Congress approved about $187 million (in 2010 dollars), which was just about $1 million less than was requested by the outgoing Clinton Administration.

President George W. Bush Administration

Bush Administration's 2001 Initiative to Increase Funding

In May 2001, President George W. Bush's National Energy Policy Development (NEPD) Group released the National Energy Policy Report. In regard to improving national energy security, that report stated:

> Energy security also requires preparing our nation for supply emergencies, and assisting low-income Americans who are most vulnerable in times of supply disruption, price spikes, and extreme weather.[24]

The report cited the cost-effectiveness of the program:

> Currently, each dollar spent on home weatherization generates $2.10 worth of energy savings over the life of the home; with additional economic, environmental, health, and safety benefits associated with the installation and resulting home improvements. Typical savings in heating bills, for a natural gas heated home, grew from about 18 percent in 1989 to 33 percent today.[25]

As one part of the plan to assist low-income people with high energy costs,[26] the report indicated that the Bush Administration was committed to increasing support for DOE's program over the long-term. Specifically, the NEPD Group recommended that:

> ... the President increase funding for the Weatherization Assistance Program by $1.2 billion over ten years. This will roughly double the spending during that period on weatherization. Consistent with that commitment, the FY2002 Budget includes a $120 million increase over 2001.[27]

As **Table A-1** and **Figure B-1** show, the Congress responded to the President's funding initiative by approving a nearly 50% increase for FY2002—raising funding from about $187 million to about $277 million, in constant 2010 dollars. The requests—and appropriations—climbed steadily for the next four years, reaching $261 million in FY2006. After that, the requests turned sharply downward and appropriations dropped slightly, standing at $232 million in FY2008.

Bush Administration's FY2009 Request to Terminate Funding

The FY2009 request observed that a 2003 assessment had rated the program as "moderately effective,"[28] and that:

> the program coordinates effectively with other related government programs in its efforts to meet interrelated Departmental goals and achieves its goals of a favorable benefit-cost ratio and other performance goals, based on internal programmatic assessments.[29]

DOE also noted that a new evaluation of the program's benefits and costs was underway. However, in a reversal of its previous requests, the George W. Bush Administration requested that funding for the Weatherization Program be terminated in FY2009. The main rationale given for program termination was that the energy efficiency technology programs operated by DOE's Office of Energy Efficiency and Renewable Energy (EERE) provided a much higher benefit-to-cost ratio than that for the Weatherization Program.[30] Specifically, the narrative for the EERE in DOE's FY2009 Congressional Budget Request stated that:

> In FY2009, Weatherization Assistance Program funds are redirected to R&D programs which deliver greater benefits. EERE's Energy Efficiency portfolio has historically provided approximately a 20 to 1 benefit to cost ratio. In comparison, Weatherization has a benefit cost ratio of 1.53 to 1.[31]

President Barack Obama Administration: Recovery Act Expands Program

Under the Obama Administration, the American Recovery and Reinvestment Act of 2009 (Recovery Act, P.L. 111-5) identified the program as one avenue to help address the recession and provided the program with a major one-time increase—$5 billion to weatherize an estimated 600,000 homes.[32] This objective aimed to help achieve the President's goal of weatherizing one million homes per year. In addition to the funding increase, some amendments to the original authorizing law were enacted. The amendments allowed more cost-effective measures to be installed in more homes. One amendment raised the ceiling on cost per dwelling from $2,500 to $6,500.[33]

Program Appeared "Shovel Ready"

The Recovery Act aimed to stimulate the U.S. economy, create jobs, and make infrastructure investments in energy and other areas. During congressional consideration of the Recovery Act, and following its enactment, the conventional wisdom was that DOE's Weatherization Program was about as close to meeting the definition of "shovel ready" as virtually any program in DOE's portfolio. Specifically, the Recovery Act weatherization effort had the following attributes: an existing programmatic infrastructure, including processes and procedures, which had been in place for many years. DOE's Inspector General observed that:

> The techniques for weatherization tasks were well known and comparatively uncomplicated, and the requisite skills were widely available; performance metrics were relatively easy to establish and understand; the potential benefits for low income citizens were easily recognized; and, the potential beneficial impact on energy conservation was obvious.[34]

With that well-entrenched program structure, there was a strong expectation that the $5 billion in Recovery Act funds would have a prompt and easily measurable impact on job creation and economic stimulation.[35]

New Program Requirements

The Recovery Act modified the program to incorporate a new labor requirement for "prevailing wage" and to embrace new training requirements.

Prevailing Wage

The Recovery Act required that state and local recipients of weatherization funds ensure that laborers be paid at least the "prevailing wage" as determined under the Davis-Bacon Act.[36] This requirement had not been previously applied to weatherization program activities. As a result, grantees lacked information on which to base wage rates. Many grantees chose not to begin work until the prevailing wage rates were formally established by the Department of Labor (DOL). Even after DOL's work was complete, additional delays occurred while grantees prepared guidance for sub-recipients on how to apply the wage rates. Thus, the delivery of Recovery Act-funded weatherization activities did not reach full momentum until after guidance was completed in October 2009.[37]

Additional Training

To ensure successful implementation of program criteria under the Recovery Act, DOE required that program personnel at state and community action agencies receive additional training. The aim of the training was to ensure that recipients understood Davis-Bacon Act requirements, new income eligibility requirements, increased allowable costs per unit, and monitoring of work performed by sub-recipients. As with other state and local activities, recession-driven budget shortfalls and staff furloughs delayed many required state training initiatives.[38]

Recession Impeded State Capacity

Recession-driven state budget shortfalls caused certain states to administer hiring freezes that applied to all employees regardless of the source of their funding, including those tasked with weatherization-related work. In some other states, progress was impacted because personnel involved with the program were subjected to significant state-wide furloughs. Further, the approval of state budgets was delayed in states such as Pennsylvania, as legislators deliberated over how to address overall budget shortfalls. Lacking staff, states were unable to perform required implementation tasks necessary to handle the large infusion of Recovery Act funding for DOE's weatherization program. Without adopted budgets, states did not have any spending authority. As a result, many states were not able to obligate or expend any weatherization program funds.

Major Implementation Delays

DOE had taken a number of proactive steps to foster timely implementation of the program. In spite of those efforts, grantees had made little progress in weatherizing homes by the beginning of 2010. As of February 2010, the one-year anniversary of the Recovery Act, only a small percentage of Recovery Act weatherization funds had been spent and few homes had actually been weatherized. Only $368.2 million (less than 8%) of the total award of $4.73 billion had been drawn by grantees. With the low spending rates, state and local grant recipients fell significantly short of goals to weatherize homes. The lack of progress by state and local grantees to implement the weatherization program funding alarmed the DOE Inspector General (IG). In early 2010, the IG found that the nation had not realized the potential economic benefits of the $5 billion in Recovery Act funds allocated to the program. In particular, the job creation impact of what many considered to be one of DOE's most "shovel ready" projects had not materialized. Further, the IG observed that modest income residents had not enjoyed the benefits of reduced energy use and better living conditions that had been promised as part of the Recovery Act weatherization effort. The IG's report found:

> Department officials worked aggressively with the states and other responsible agencies to mitigate these challenges.... However, as a practical matter, program challenges, such as those identified in this report, placed the Recovery Act-funded Weatherization Program "on hold" for up to nine months.[39]

Thus, the Recovery Act goals proved to be much more difficult to achieve than originally envisioned. The IG report noted:

> The results of our review confirmed that as straight forward as the program may have seemed, and despite the best efforts of the Department, any program with so many moving parts was extraordinarily difficult to synchronize. In this case, program execution depended on the ability of the Federal government (multiple agencies, in fact), state government, grant sub-recipients and weatherization contractors, working within the existing Federal and state regulatory guidelines, to respond to a rapid and overwhelming increase in funding. Further, anticipated stimulus impact was affected by certain conditions and events clearly outside of Departmental control including state budget difficulties; availability of trained and experienced program staff; and, meaningful changes in regulatory requirements.[40]

Thus, despite the assumption that the program was "shovel ready," the IG uncovered several key administrative and intergovernmental barriers to delivery of Recovery Act funded weatherization services.

PERFORMANCE ASSESSMENTS AND EVALUATION STUDIES OF COST-EFFECTIVENESS

This part of the report reviews a selection of key studies that examined Weatherization Program management processes, performance, and economic (energy and non-energy) impacts. Performance assessments and evaluation studies are related activities within a family of research on performance measurement and program evaluation. Both activities support the Government Performance and Results Act (GPRA) and the Office of Management and Budget's (OMB's) concerns for federal program management. In general, a performance assessment employs anecdotal data that is collected over a shorter time-frame, with a focus on management responsibilities and budget processes. In contrast, an evaluation study employs survey research methods to gather comprehensive data that is collected over a longer time-frame, with a focus on revealing the net impacts and cost-effectiveness of program operations.[41]

Background

GPRA Performance Assessment Requirements

The Government Performance and Results Act of 1993 (GPRA, P.L. 103-62) established requirements for federal agencies to conduct regular performance assessments based on general methods of management by objectives and strategic planning. The law directs OMB to lead GPRA implementation,[42] in cooperation with federal agencies. OMB describes this responsibility in Part 6 of its Circular A-11:

> GPRA provides the foundation for performance planning, reporting, and budgeting for Federal agencies. GPRA requires agencies to prepare strategic plans, related performance budgets/annual performance plans, and annual performance reports (31 U.S.C. § 1115). The legal requirements for an annual performance plan are met by a performance budget. The annual performance report requirement (APR) will be

fulfilled by either the annual performance and accountability report (PAR) or by the congressional budget justification for agencies that choose to produce a separate annual financial report (AFR) and APR.[43]

Among more specific GPRA purposes, Circular A-11 identifies a key GPRA purpose to:

> Improve Congressional decision-making by providing more objective information on achieving statutory objectives, and on the relative effectiveness and efficiency of Federal programs and spending....[44]

Over the years, DOE has integrated its responses to GPRA requirements into its strategic plan, annual performance report,[45] annual performance and accountability report,[46] and annual budget request. DOE notes that it has, since 2002, been working with OMB to assess its programs with the Performance Assessment Rating Tool (PART).[47] Evidently, PART had become the main vehicle for reporting of GPRA assessment requirements. PART is described in greater detail in the next section of the report.

Circular A-11 defines program assessment:

> A determination, through objective measurement and systematic analysis, of the manner and extent to which Federal programs achieve intended objectives.[48]

It also defines program evaluation:

> Program and practice evaluation is an assessment, through objective measurement and systematic analysis, of the manner and extent to which programs or practices achieve intended results. Program and practice evaluations should be performed with sufficient scope, quality, and independence. Although agencies may cite rigorous evaluations commissioned independently by organizations such as the Government Accountability Office, Office of the Inspector General, or other groups, these evaluations should not completely supplant rigorous evaluations commissioned by the agencies themselves.[49]

DOE's Approach to Assessment and Evaluation

DOE's Weatherization Program has been assessed through a combination of assessment and evaluation methods. DOE's GPRA-driven program assessment tools—strategic plan, annual performance plan, performance and accountability report, and performance budget—tend to be quite broad, focused on the larger agency missions of R&D, science, and defense. Those documents tend to mention the program incidentally, mainly citing achievements in terms of annual number of units weatherized. For a more in-depth examination of program operations, DOE has a history of directing the Oak Ridge National Laboratory (ORNL) to conduct evaluation impact studies.

Choice of Studies for Review

In 1993, DOE published a major evaluation report, the first "scientific" study of program performance, impact, and cost-effectiveness. That report established impact evaluation studies as a key feedback component that promoted program design changes and improvements. However, its cost, complexity, and lag time led DOE to subsequently rely on aggregations of available state-level evaluations.

Also, in 1993, GPRA stimulated the emergence of a second track of less in-depth performance assessments that had a shorter-term budget focus.

Both tracks of analysis—performance assessments and evaluation studies—have the same general goals of providing feedback that can be used to improve program operations. So both types of studies were selected for review.

Selected Performance Assessments

2003 IG Performance Audit

A performance audit is an assessment tool used to examine the performance and management of a program against objective criteria. It is designed to facilitate oversight and may serve several objectives, including the assessment of program effectiveness and results.[50] From late 2002 through early 2003, DOE's Office of the Inspector General (IG) conducted a performance audit of the program. The audit period included Program Year 2000 (PY2000), PY2001, and planned activities for PY2002.[51] The total DOE budget for the audit review period was $518 million. The purpose of the audit was to "determine whether the program was properly administered and was achieving its goals."[52] The audit was conducted in accordance with Government Auditing Standards for performance audits.[53]

The IG observed that the program has a long-established structure for transferring funds to state and local agencies and that improvements over the years had made use of the funds more efficient and effective.[54] As further background, the IG observed that, in addition to DOE funds, state and local agencies also obtain funding for weatherization from the Department of Health and Human Services' (HHS) Low-Income Home Energy Assistance Program (LIHEAP) and other programs funded by utilities, states, and other sources.

Local agencies report to DOE annually through state weatherization offices on expenditures, number of units completed, and other performance-related measures. For PY2001, the IG noted that "about 900 agencies received funding that ranged from as low as a few thousand dollars to as high as $4 million."[55] The IG found two local government reporting issues: administrative costs and number of households weatherized.

Regarding the first concern, it identified instances where local agencies reported administrative-type expenses as program operating costs:

> Specifically, we observed that certain organizations inappropriately charged expenses such as administrative staff, office rent, and administrative supplies as direct program costs and thus understated total administrative costs. Public laws and federal regulations limit the amount of weatherization grant funds that may be used for administrative purposes. If local agencies continue to under report administrative costs, states could ultimately exceed statutory limitations for administrative expenses over the life of the grant.[56]

DOE advised the IG that it is often difficult for local agencies to operate within the limits for administrative costs. The IG concluded that the DOE Program Office should work more closely with states and local agencies to ensure that administrative costs are minimized and that the costs incurred are reported accurately and consistently.

Regarding the concern about local agency reporting on the number of households weatherized, the IG found that:

> ... data regarding the number of households weatherized was not reported on a consistent basis. For example, data reported by some states related strictly to the number of homes weatherized using Department funds. Other states, however, combined the results of weatherization efforts funded by the HHS LIHEAP with those completed with Departmental funds. Merging performance data for such states distorts program results and could make it appear program efficiencies and energy savings are greater than that actually achieved with available funding.[57]

A recent OMB program assessment found that DOE had addressed the IG's concern about distorted efficiency (benefit-cost) measures:

> Average cost per home employed in the calculation of benefit/cost ratios now reflects total costs (including non-DOE funds) expended per unit in the states whose evaluations were used in the energy savings estimates. This substantially raises the per-unit investment from a DOE-only level making the benefit/cost ratio more conservative.[58]

2003 OMB Assessment with PART Method

PART Background

The Performance Assessment Rating Tool (PART) was developed by the Office of Management and Budget (OMB) in 2002 as a key component for implementing GPRA and the President's Management Agenda (PMA),[59] particularly the Budget and Performance Integration initiative.[60] PART was designed to assess program planning, management, and performance against quantitative, outcome-oriented goals. It can inform funding and management decisions aimed at improving program effectiveness.[61] As a diagnostic instrument for evaluating efficiency and effectiveness, PART aims to help managers identify, anticipate, and rectify performance problems.

PART was designed to provide a basis for DOE and OMB to agree upon meaningful annual and long-term targets. The PART assessment process aims to unify annual performance targets and goals with long-term goals. PART was designed to be an iterative process, capable of tracking the evolution of program performance over time through periodic reassessments. OMB's recommendations to foster program improvement are central to the PART process. Program offices track actions taken to implement PART recommendations and report those actions to OMB semi-annually.

According to OMB, the ongoing cycle of reviewing and implementing PART recommendations, coupled with the use of performance data from assessments and periodic reassessments, signify the perception of PART as an integral process for planning and budget decision-making, as distinguished from a set of one-time program evaluations.[62]

PART assessments help inform budget decisions and identify actions to improve results. Agencies are held accountable for implementing PART follow-up actions and working toward continual improvements in performance. PART asks a series of questions that cover four areas:

- program purpose and design,
- performance measurement, evaluation, and strategic planning,
- program management, and
- program results.

To earn a high PART rating, a program must use performance assessment to manage, justify its resource requests on expected (projected) performance, and continually improve efficiency. All of those activities are goals of the Budget and Performance Integration Initiative. The PART performance rating scale is shown below:

PART Report on DOE Weatherization Program

The summary of a PART report on the DOE Weatherization Program found that it was performing at the "moderately effective" level.[63] The assessment defines the moderately effective rating as follows:

> In general, a program rated Moderately Effective has set ambitious goals and is well-managed. Moderately Effective programs likely need to improve their efficiency or address other problems in the programs' design or management in order to achieve better results.[64]

The PART assessment cited three findings to explain the "moderately effective" rating. First, it noted that the program met its annual performance target for the number of homes weatherized. Second, OMB stated that the program lacked an assessment of performance that was current, comprehensive, and independent. It noted that the program reported on "internal assessments" that showed a favorable benefit-cost ratio. However, OMB found that those assessments relied in part on old data and were conducted by Oak Ridge National Lab (ORNL), which it says is not an "independent" source.[65] Third, OMB noted that the DOE Inspector General identified issues with the way state and local agencies report on program management.[66] DOE responded that it was conducting an independent analysis of the cost-effectiveness of the program and addressing the IG's audit recommendations.

The detailed PART assessment report presented trend data for key performance measures, including number of homes weatherized, cost per home, and the overall benefit-cost ratio.[67] Also, it described and rated strategic program features, such as goals, objectives, design, targets, and budgeting.

Table 2. PART Rating Scale (the numerical rating represents the total point score on the PART questionnaire)

Rating	Range
Effective	85 -100
Moderately Effective	70 -84
Adequate	50 -69
Ineffective	0 - 49

Source: National Institutes of Health. GPRA and PART. http://sourcebook.od.nih.gov/SD%20Orientation/PART.pdf.

The PART Assessment defined the benefit-cost ratio as a long-term measure of program efficiency.[68] OMB stated that the benefit-cost ratio represents the value of energy saved,[69] divided by program costs.[70] The ratio depended in part on estimated long-term energy prices at the time of the assessment and on an average energy savings per household of 29.1 million British thermal units (MBtu).[71] PART reported that estimates of the benefit-cost ratio for energy savings under various price scenarios ranged from 1.19 to values greater than 2.00,[72] never measuring less than 1.00.[73]

PART assessment question number 4.3 asked:

> Does the program demonstrate improved efficiencies or cost effectiveness in achieving program performance goals each year?

The response was:

> Benefit-cost ratio rose from 1.06 in 1989 to 1.79 in 1996, and then declined to 1.51 and 1.30 in 1999 and 2002, respectively. These estimates depend largely on EIA estimated long-term energy prices.

PART question number 2.6, on strategic planning, asked:

> Are independent and quality evaluations of sufficient scope and quality conducted on a regular basis or as needed to support program improvements and evaluate effectiveness and relevance to the problem, interest, or need?[74]

The response touched upon several aspects of the evaluation strategy, including frequency of studies, measurement of non-energy benefits, and independence of the evaluator. It stated:

> The program does not conduct annual evaluations on a national basis because of the high cost of such evaluation and the limited amount of change that occurs in program activities from year to year. The program has contracted with Oak Ridge National Laboratory (ORNL) to devise evaluation methodologies and report periodically on program results based on state grantee-level performance evaluation. ORNL has also conducted selective evaluation activities designed to inform program management of performance characteristics in areas in which the program performance has been below average (hot climate zones) or in areas in which there has been growing strategic program interest but little evaluation data. The latter includes base load electric measures as well as non-energy benefits. To assure independence, the program should consider using an alternative contractor in future assessments, or at least having future ORNL reports assessed by a third party.[75]

As evidence of its evaluation activities, DOE's response included references to several ORNL program metaevaluations and other evaluation studies.[76] Some key ORNL studies are described in the next section.

2011 GAO Performance Audit of Recovery Act Funding

The Government Accountability Office (GAO) defines a performance audit broadly:

> Performance audits provide objective analysis so that management and those charged with governance and oversight can use the information to improve program performance and operations, reduce costs, facilitate decision making by parties with responsibility to oversee or initiate corrective action, and contribute to public accountability.[77]

According to GAO, a performance audit may embrace a wide variety of objectives, including:

- an analysis of the "relative cost-effectiveness of a program or activity," and/or
- a determination of the "current status or condition of program operations or progress in implementing legislative requirements."[78]

Thus, by GAO's definition, a performance audit may include activities with the same purposes as those of an impact and/or process evaluation study.

The Recovery Act directed GAO to conduct bimonthly reviews and reporting on selected states and localities' use of the funds it provided, including the funds for the DOE Weatherization Program.[79] GAO describes these bimonthly review as performance audits. In addition to the bimonthly audits, GAO's response to the directive has included one performance audit that covered only the Weatherization Program. For example, in December 2009, GAO published a performance audit report on Recovery Act funds used by states and localities.[80]

The DOE Weatherization Program was one of many programs covered. For the weatherization component, GAO covered 16 states and 24 localities. GAO found that, as of the end of FY2009, states had outlaid about 2% ($113 million) of weatherization funds and had completed about 1% (7,300) of the targeted number of homes to be weatherized.[81] Many contracts between states and local agencies had been delayed due to concerns about staff capacity and new labor requirements.[82] As another example, in May 2010, GAO published another audit report on Recovery Act funds that covered DOE Weatherization Program funds.[83]

As of the mid-point in FY2010, states had outlaid about 13% ($659 million) of weatherization funds and had completed about 13% of the targeted number of homes to be weatherized. GAO made several observations about the status of program implementation and offered some recommendations for program improvement.[84] In January 2011, the GAO initiated a performance audit focused solely on Recovery Act funding for the DOE Weatherization Program.[85] The report was released in late December 2011.[86] The study aimed to examine: (1) status of funds, (2) implementation challenges facing recipients, (3) achievement of energy and cost savings, and (4) changes in the quality of employment data reported by recipients.

To address the four objectives, GAO conducted a web survey of all 58 recipients and received 55 responses.[87] GAO also interviewed officials from DOE and selected national associations, conducted site visits in seven states, and conducted telephone interviews with two other state agencies.[88] Selected findings, as of the end of FY2011, included: (1) recipients had spent about $3.46 billion (73%) of the $4.75 billion allocated,[89] (2) recipients reported that implementation challenges were declining,[90] (3) the average cost per home was about $4,900,[91] about 563,000 homes had been weatherized, and DOE projected it would exceed the original target of 607,000 homes by the deadline set for the end of March 2012;[92] and (4) GAO found that the quality of employed data reported by recipients had improved.[93]

Selected Evaluation Studies

Several DOE reports, prepared primarily by staff of the Oak Ridge National Laboratory (ORNL), have assessed the energy cost savings and non energy benefits of the DOE Weatherization Program.[94] In particular, a comprehensive evaluation of the PY1989 program was published in 1993.[95]

1993 National Evaluation Report

In 1993, DOE issued a report entitled *National Impacts of the Weatherization Assistance Program in Single Family and Small Multifamily Dwellings*.[96] The reporting project was an exhaustive five-year, first-of-its-kind, comprehensive evaluation study of the energy savings and cost-effectiveness of the program during PY1989.[97] A letter accompanying the report noted that the program is "the nation's largest residential conservation program, and one of its oldest." The letter captures the highlights:

> The report evaluates the national and regional energy savings and cost effectiveness of the program in single-family dwellings, mobile homes, and small (2-unit to 4-unit) multifamily dwellings. It is based upon a representative national sample of nearly 15,000 dwellings weatherized by 368 local weatherization agencies, and a control group of homes waiting to be weatherized by the same agencies.[98]

The report estimated an overall program benefit-cost ratio of 1.09 for the entire program.[99] Adding estimates of non-energy benefits—such as comfort, employment, reduced environmental impacts, and preservation of affordable housing—yielded a societal benefit-cost ratio of 1.72.

Three Metaevaluations (1997, 1999, 2005)

After the 1993 National Evaluation report, DOE relied upon "metaevaluations"[100] of state evaluation studies as the main tool to assess program energy savings and cost effectiveness. All of the metaevaluations were conducted by ORNL. Compared with the 1993 National Evaluation, metaevaluations became a low-cost way for DOE to update energy savings and cost-effectivness. This section reviews three key metaevaluations, covering program operations from 1990 through 2005.

1997 Metaevaluation (Covers 1990-1996)

DOE asked ORNL to prepare the 1997 metaevalution.[101] The study involved locating, assembling, and summarizing the results of all state-level evaluations that had become available since 1990. ORNL found 17 state-level evaluation studies from 15 states that it used to cover program operations during the period from 1990 through 1995.[102] Energy savings were expected to be higher than those found in the 1993 National Evaluation, which was based on 1989 data. Typical savings ranged from 18% to 24%, which was greater than the range from 12% to 16% found in the 1993 National Evaluation.[103] Key reasons for the expected improvement included some advances in weatherization procedures, such as the use of advanced audits and blower-door directed air sealing.[104] The study concluded that program performance had improved significantly over that seven-year period.[105]

The 1997 metaevaluation study used three perspectives to estimate cost-effectiveness. The program perspective compared energy benefits to total program costs. The installation perspective compared energy benefits to installation costs. The societal perspective compared the summed value of energy and non-energy benefits to total costs.[106] ORNL reported that the metaevaluation calculations employed the same procedures and assumptions used in the 1993 National Evaluation. The values of 1.79 (program perspective), 2.39 (installation perspective), and 2.40 (societal perspective), respectively, are shown in **Table 3**.[107]

Table 3. Benefit-Cost Ratios for the 1993 National Evaluation, Three Metaevaluations (1997, 1999, and 2005), and the 2010 Technical Memorandum (Three benefit-cost perspectives shown)

Study	Program Perspective	Installation Perspective	Societal Perspective
National Evaluation (1989)	1.06	1.58	1.61
Metaevaluation (1990-1996)	1.79	2.39	2.40
Metaevaluation (1996-1998)	1.51	2.02	2.12
Metaevaluation (1993-2005)	1.34	—	2.53
Technical Memorandum	1.80		2.51

Sources: ORNL/CON-326; ORNL/CON-435; ORNL/CON-467; RNL/CON-493; ORNL/TM-2010/66.

The study found that the synthesis of state-level evaluations offered a reasonable characterization of national program performance:

> Although national level evaluation efforts are sometimes needed to definitively demonstrate program performance, reviews of state-level evaluations provide useful, and inexpensive, benchmarks of progress during the years between such large-scale national assessments.[108]

1999 Metaevaluation (Covers 1996-1998)

As a follow-up to the 1996 metaevaluation, ORNL performed another metaevaluation, which was published in 1999. The report synthesized results from 10 individual evaluation studies of state weatherization efforts that were completed between April 1996 and September 1998.[109]

The stated objectives of the 1999 study were (1) to identify average energy savings of households in the states that provided information; (2) to identify key variables that explain the magnitude of those weatherization-induced savings; and (3) to use the findings from the state studies to estimate average household energy savings that could be expected nationwide.[110]

The study reported that program-induced savings continued at a higher rate than those found in the 1993 national evaluation.[111] However, the study found no dramatic changes were made in the structure or practices of the program during the period after the 1996 study.[112] Thus, no major increases in savings were expected above those found in the 1996 study. To make the benefit-cost ratios in the 1999 study comparable to those in the 1996 metaevaluation and the 1993 national evaluation, ORNL used the same assumptions and procedures. In particular, the average measure lifetime was assumed to be 20 years and the discount rate used was 4.7%. Cost-effectiveness was calculated for the weatherization program nationwide. As in past evaluations, three different benefit-cost perspectives (program, installation, and societal) were examined. Because higher average national energy savings were estimated in the 1996 and 1999 metaevaluations, the benefit-cost ratios for those years were higher than the ones reported in the 1993 National Evaluation. The benefit-cost ratios for the three perspectives are shown in **Table 3**.

2005 Metaevaluation (Covers 1993-2005)

The most recent metaevaluation, published in 2005, estimated average household energy savings at 30.5 million British thermal units (MBtu) per year with a benefit/cost ratio from the program perspective estimated at 1.34 with 2003 prices. Costs were based on those derived from the state evaluations themselves.[113]

Like the previous metaevaluations, ORNL undertook this one to update the findings from the 1993 national evaluation.[114] This metaevaluation was based on data from 38 evaluation studies of weatherization programs in 19 states,[115] published between 1993 and 2005.[116]

The overall method of the 2005 metaevaluation paralleled that of the previous metaevaluations, but ORNL describes some improvements:

1. Expanded Range of Time and Geography. The overall similarity of results across the various studies led ORNL to include findings from a 2003 study that covered all post-1992 state-level studies. ORNL found that this approach increased sample size, improved the ability to cover all major U.S. climate regions, and added to the statistical rigor of the results.
2. States as the Unit of Analysis. ORNL aggregated data from the 38 studies by state. This aspect contrasts with previous metaevaluations, which had treated multiple studies of the same state as separate empirical observations.[117]
3. Studies Combined into a Single State Value. For each relevant variable (e.g., energy savings per gas-heated household), a single average value was used to represent the entire state. ORNL says this procedure prevents overall study results from being skewed due to the presence of multiple studies with similar results in a single state.
4. Weighted Average Used to Account for Multiple Studies. Where there were multiple studies, the average values computed in each study were used to determine a weighted average for each variable, with the weighting done by sample size.[118]

Under this revised approach, those states that represent a larger part of the national program contribute more heavily to the analysis and findings. This is appropriate because the purpose of the study is to estimate energy savings nationwide.[119] **Table 3** shows the benefit-cost values in the various ORNL metaevaluations. The similarity among the values is not surprising because those meta analyses used data from many of the same state-level studies.[120] ORNL states that all three of the metaevaluations, including the 2005 study, focused primarily on energy savings in homes heated by natural gas because the large majority of state-level studies addressed that fuel.[121] ORNL found a statistically significant difference between the natural gas savings per household in the 2005 metaevaluation and the savings reported in the 1993 national evaluation.[122] However, the 2005 study noted that the other two-thirds of the states were not included in the metaevaluation, and thus did not provide data for the study, which raises the possibility that the sample examined may not fully represent the nation as a whole.[123] ORNL found that actual benefit-cost values are likely to be higher than those reported for natural gas savings, as shown in **Table 3**. This was due to the fact that reported expenditures also included the cost of installing measures to reduce baseload electricity use, but only the benefits of natural gas savings were used in those benefit-cost calculations. Overall, the study found significant program improvements since 1989 (as reported in the 1993 national evaluation) and a generally higher benefit-cost ratio. However, it also acknowledged the shortcomings of the metaevaluation method and called for a new national evaluation:

> While the metaevaluations performed over the last decade have consistently shown higher natural gas savings per household than those reported in the national evaluation, there is a need to corroborate those findings through a rigorous examination of Weatherization Program efforts nationwide. Even the current metaevaluation is based on studies performed in only a third of the states, and those may not be fully representative of the entire Weatherization Program. Also, the value for preweatherization energy consumption, which is a major input for the calculation of national savings, is based on 1989 data. In addition, while state-level evaluations have put a strong emphasis on gas-heated houses, few studies have been conducted on electrically heated dwellings. And it is important to note that the biggest recent change to the Weatherization Program – the addition of baseload measures such as highly efficient refrigerators, water heaters, and light bulbs – has barely been addressed by state-level studies. For all these reasons, there is a strong need for a new national evaluation to thoroughly explore the current operations and achievements of the Weatherization Program across the entire nation.[124]

DOE has engaged APPRISE, a nonprofit analytical organization, to conduct the second national evaluation. The status of plans for that evaluation are described in the section below, entitled "Evaluation Plan for the Regular Program."

2010 Technical Memorandum (FY2010 Projected Data)

The Recovery Act provided a large infusion of funding and prescribed major new requirements for the program. In response to concerns about these major changes, ORNL prepared a technical memorandum to update estimates from the metaevaluations. [125] The memo only attempted to make preliminary projections based on an audit; it did not attempt to generate empirical data from an impact evaluation study. Specifically, the memo recounted results from an engineering modeling approach that used the National Energy Audit Tool (NEAT) to estimate the performance of the program by using typical homes in each state.[126] These homes varied by building type, primary heating fuel and prices, and local weather conditions. ORNL employed the Weatherization Assistant residential audit package to estimate annual energy savings under new parameters set by the Recovery Act.[127] The estimated annual savings for heating and cooling projected for FY2010 were 29.0 MBtu per household.[128] In comparison, the 2005 metaevaluation estimated savings at 30.5 MBtu over the period from 1993 through early 2005.[129] The Recovery Act raised the average per-unit investment cost ceiling from $2,500 to $6,500. The likely increase in investment per unit was expected to increase total energy savings. However, the higher funding level was accompanied by a shift in formula allocation that allowed proportionally more funding for states in warmer climates.[130] ORNL projected a 2010 benefit-cost ratio of 1.80 (from the program perspective). Most of the increase relative to the 2005 benefit-cost ratio of 1.54 was attributed to the change in energy prices from 2003 to 2010. Costs were based on the Recovery Act ceiling of $6,500 for average costs. However, the ceiling for cost-effective expenditures varied from state to state, based on the requirement that the savings-to-investment ratio (SIR)[131] reach one (1.0) or greater for each of three factors: the last measure installed, energy intensity, and local energy prices.[132] Traditionally, the program benefit-cost ratio has been expressed in terms of estimated national energy cost savings per dollar of investment for homes heated with natural gas, with cost savings based on national average residential natural gas prices. ORNL's estimates in the memo differ somewhat, due to reliance on state-by-state energy savings and investment numbers used to generate the energy savings estimates. This aspect allowed the dollar savings estimates to vary by region, by fuel type, by housing type, and with the adjusted average cost ceiling provided under the Recovery Act. The projected 2010 societal benefit/cost ratio (all benefits divided by all costs) is 2.51. This is almost identical to the societal benefit/cost ratio of 2.53 computed in the 2005 metaevaluation.[133] ONRL reflected on the limits of the data used to make estimates, noting again the need for a new national evaluation:

> The methods used here are more complex and hopefully better at reflecting current reality than those used in previous years. Nonetheless, one needs to keep in mind the limitations of available input data that constrain an analysis such as this. There is no nationally available data on energy-related housing characteristics of weatherized households, nor is there data on the variations in the cost of measures installed from one locale to another. There is also no way of knowing at this point the exact characteristics of the housing stock and housing types that will actually get weatherized with the greatly expanded revenues available. Making these estimates more precise requires populating the methodology with more accurate data that will flow from the National Evaluation effort.[134]

ONRL noted that the memo's estimates were derived from a different methodology than that used in the metaevaluations. There are several reasons that a change in method was needed. First, the metaevaluation estimates are dated and do not reflect recent changes in program operations that materially impact household savings. These include a major change in the program's average cost ceiling, from $2,500 to $6,500, an expansion in allowable measures to include electricity measures such as refrigerator replacement and lighting changeout, and a major increase in program funding impacting the allocation of resources among different regions and climate zones. Second, the 2005 metaevaluation results describe only homes heated with natural gas and do not reflect the diversity of heating fuels used in treated homes nor do the results reflect potential cooling savings. The method used for the metaevaluations reflected only savings in single-family homes and those savings were never adjusted for variations in the treated housing stock. Finally, the estimates in the metaevaluations were based on national average energy prices and were not varied to reflect the diversity of energy prices weighted by the location of the weatherization work being performed.[135] ONRL states that the methodology used to prepare statistics for the memo corrects for many of the above-noted deficiencies. However, it acknowledges that the findings do not present a statistically valid representation of the program's performance and results:

> There are too many assumptions and uncertainties incorporated in it to allow that to be the case. Much of this is the result of a lack of up-to-date information on program operations, particularly regarding measures installed and their cost as well as the energy-related characteristics of the homes weatherized. Rather, the results should be treated as the best currently available estimate that can serve until more rigorous results are provided by the new National Evaluation.[136]

CURRENT PLANS FOR ASSESSMENT AND EVALUATION STUDIES

Three activities are underway to assess and evaluate DOE's Weatherization program: a DOE (Deloitte) strategic assessment, a DOE (APPRISE) impact study for PY2007 through PY2008, and a DOE (APPRISE) impact evaluation for the Recovery Act period of PY2009 through

PY2011. **Table 4** provides a brief overview of the goals and target dates for the three activities. Descriptions of the activities follow.

Plan for a Strategic Assessment (Deloitte)

In November 2008, DOE's Weatherization Assistance Program Office announced that it had contracted with Deloitte Limited Liability Partnership (LLP) to perform a strategic assessment of the program.[137] Deloitte is tasked with conducting a fundamental analysis of the program's objectives, impact metrics, market delivery vehicles, and finance mechanisms. The study aims to identify fundamental improvements in program design and delivery, in contrast to the more traditional evaluation of program benefits and costs that ORNL has conducted in the past.[138]

Table 4. DOE Assessment and Evaluation Activities Underway (As of December 2011)

	Strategic Assessment	Second National Retrospective Evaluation	Recovery Act Retrospective Evaluation
Organization	Deloitte	ORNL-APPRISE	ORNL-APPRISE
Process Evaluation	yes	yes	yes
Impact Evaluation	no	yes	yes
Period Covered	—	2007 through 2008	2009 through 2011
Focus	program design and measurement tools	comprehensive energy and non-energy impacts, benefit-cost	comprehensive energy and non-energy impacts, benefit-cost
Target Date for Completion/Report	No date published	Fall 2012	Early 2014

Source: DOE, various references.

Notes: The only published target dates for completion were obtained from ORNL, *Evaluation of the National Weatherization Assistance Program during Program Years 2009-2011 (American Reinvestment and Recovery Act Period)*, ORNL/TM-2011/87, May 20, 2011 (draft date), pp. 1 and 123.

Evaluation Plan for Regular Program (APPRISE, 2007-2008 Data)

DOE acknowledges that OMB's PART review and the IG's report have highlighted the need for an updated comprehensive evaluation of the program to ensure that objectives are being met and that estimates of energy savings, bill reductions, program costs, and program benefits are valid. Since the first national evaluation was published in 1993, there have been several changes made to program policies, procedures, technologies, and techniques. DOE notes that it is important to assess how well these changes have worked. For example, there is a need to assess how well new policies—such as whole-house weatherization—are impacting the program.[139] In response to the growing need for updated and valid program data, the program moved forward in FY2005 to begin planning another national evaluation. With a committee comprised of state and local weatherization staff, the program office started planning the evaluation and identifying areas where the program might be improved. DOE states that the 1993 national evaluation brought about major program changes such as computerized audits, less use of door and window measures, and increased emphasis on hot climate opportunities. DOE anticipates that the second national evaluation will have a similar effect on future program policy and performance.[140] In February 2007, ORNL released a plan for a second national retrospective evaluation, which would use data collected during PY2006.[141] ORNL says that the program has been changed by findings from the 1993 study, a strategic planning process, and other initiatives that had caused the program to change significantly. The program has incorporated new funding sources, management principles, audit procedures, and energy efficiency measures.[142] ORNL is supervising a competitively selected independent contractor team to conduct the evaluation. The team is led by the Applied Public Policy Research Institute for Study and Evaluation (APPRISE).[143] The study will assess program performance for PY2007 and PY2008, the period immediately preceding the Recovery Act. Performance measures will include program costs and benefits. In order to conduct a comprehensive savings analysis, the design includes collection of a full year of post-weatherization billing data.[144] The evaluation is expected to be completed by Fall 2012.[145]

Design for the APPRISE Second National Retrospective (Impact) Evaluation

DOE's central evaluation question is: How much energy did the weatherization program save in 2007 and 2008? DOE plans to use well-known analytical approaches to answer this question. Electricity and natural gas billing histories will be collected pre- and post-weatherization for a sample of weatherized homes (the treatment group) and a sample of comparable homes that were not weatherized (the comparison group).[146] The comparison group for the 2007 treatment group will be homes weatherized in 2008. The comparison group for the 2008 treatment group will be homes weatherized in 2009. A national sample of 400 local weatherization agencies will be selected to provide information on about one-third of their weatherized homes. Billing history data will be normalized using three different analytic methods, including the Princeton Scorekeeping Method (PRISM).[147]

DOE will also address an important complementary evaluation question: How cost-effective were those energy savings? In principle, all measures should be cost-effective, because each one is required to meet the savings-investment ratio (SIR) test. Evaluations are conducted to compare expectations with measured values. DOE says it will collect cost information associated with each weatherized home included in the treatment sample. Total projected energy cost savings over the lifetime of the measures will be divided by total costs to estimate a benefit-cost ratio for the program. The study will attempt to quantify nonenergy benefits, which will be sorted into three categories—utility benefits, occupant benefits, and societal benefits. Utility providers benefit because weatherization reduces arrearages and service shutoffs. Occupants benefit because weatherization makes homes more comfortable, healthier, safer, and more valuable. Society benefits because weatherization curbs greenhouse gas emissions, reduces other forms of pollution, and conserves water. Further, it increases local employment and it increases local economic activity caused by the multiplier effect. [148]

Design for APPRISE Process Evaluation

DOE intends that the Second National Retrospective Evaluation also examine program processes. To do that, it plans to survey all program grantees and subgrantees to collect data about program operations, training and client education activities, and quality assurance procedures. That survey data will be used to provide a program snapshot for PY2008. Case studies and a field study will be incorporated into the process evaluation design to furnish additional insights on program implementation.[149]

Evaluation Plan for Recovery Act (APPRISE, 2009-2011 Data)

The Recovery Act committed $5 billion over two years to an expanded weatherization program. This large sum greatly heightened interest in the program, the population served, its energy and cost savings, and its cost-effectiveness. Solid answers to many of the questions about the program and its performance require a comprehensive evaluation. DOE is funding an evaluation by APPRISE[150] in order to ascertain program effectiveness and to improve program performance.[151] The evaluation project has a budget of $19 million. GAO reviewed the plan's design for evaluating energy cost-effectiveness and found it methodically sound.[152] DOE observes that the weatherization program has changed considerably. In particular, it notes that there are four key differences between the pre-Recovery Act program and the post-Recovery Act program.[153] First, to expend $5 billion and to increase weatherization activity by 300%, a greatly expanded weatherization workforce was recruited, trained, organized, and sent into the field. To support that expansion, the percentage of program funds allowed for training and technical assistance was raised from 10% to 20%. Second, all states and U.S. territories received large funding increases, and some states and local agencies struggled with weatherization budgets several times larger than previous ones. Faced with that major program expansion, many states and local agencies chose to implement innovations in program delivery and management. Third, DOE also has made provisions to set aside substantial sums to support innovations in program funding and design. Fourth, to accommodate the Recovery Act funding expansion, major changes were incorporated into the program:

- The definition of the threshold for a low- income household was raised from150% to 200% of the federal poverty income guidelines;
- The ceiling on average cost per unit—that is, the average amount of money that grantees can spend to weatherize their homes—was increased from $2,500 to$6,500; and
- Wages for weatherization workers were subjected to Davis-Bacon Act prevailing wage requirements for the first time.

DOE/ORNL says that many features of the 2009-2011 evaluation will parallel those of the retrospective evaluation for 2007-2008, as described in the previous section. Analyses of energy savings and cost-effectiveness will be similar, although billing histories may be collected to also assess the effects of certain program innovations. The main difference between the retrospective evaluation of 2007-2008 and the evaluation of 2009-2011 will be the process evaluation component. DOE says that funding and associated provisions have had a significant impact on the weatherization community and program operations. Thus, it plans two special process evaluation studies as part of the Recovery Act evaluation project. One would focus on Davis-Bacon requirements and the other would address changes in the composition of the national weatherization community. In May 2011, ORNL published a plan for the Recovery Act evaluation that includes details about both the impact and the process components of the assessment.[154] The schedule for the evaluation activities covers the period from the second quarter of calendar year (CY) 2011 through the first quarter of CY2014.[155]

Impacts of Davis Bacon Requirement

Congress directed that weatherization projects funded with Recovery Act appropriations must follow Davis-Bacon labor rules.[156] The Department of Labor is responsible for identifying prevailing wages in the construction industry. These wages are identified for a set of construction industry jobs and are estimated for each county in the United States. Two possible questions DOE has identified for this part of the evaluation research are:

> (1) What were the actual, monetary administrative costs for complying with Davis-Bacon?
>
> (2) Overall, how did application of Davis-Bacon influence the cost-effectiveness of WAP?

Impacts on Administration, Training, and Employment

DOE observes that the large flow of Recovery Act funds into low-income weatherization changed the national weatherization network. The weatherization labor force grew larger and new stakeholders were drawn into the network. The increased funding had both positive and negative effects on long-standing leveraging relationships. DOE is considering several possible research questions. Four of the questions are:

> (1) How have relationships between state weatherization offices and local weatherization agencies changed?
>
> (2) Did the Recovery Act change the way that local agencies use contracts to procure weatherization services?[157]
>
> (3) What approaches did local agencies and/or contractors use to recruit and train new qualified, reliable, and trustworthy weatherization crew members, and how effective were these approaches?
>
> (4) Did the expanded weatherization workforce find work opportunities in the energy efficiency field outside of the DOE weatherization program?

Other Expected Results

The evaluation design is expected to produce a variety of new information and findings about the program. Basic results are expected to include estimated energy savings for 2007 through 2011, along with estimates of cost-effectiveness and non-energy benefits. The design aims to uncover insights into energy savings attributable to particular measures and the strengths and weaknesses of computer audits, versus priority lists, as a way to select appropriate weatherization measures. Also, the design is expected to produce program operational data for two levels of average home investments ($2,500 and $6,500) and two levels of home eligibility (150% and 200% of the poverty level). Findings are expected to yield some indication of the impact on program benefits attributable to changing the federal grant formula in a way that distributes more funds to states in warmer climates.[158]

ISSUE OF EVALUATION INDEPENDENCE AND OBJECTIVITY

The OMB Inspector General has criticized DOE for depending almost exclusively on "in-house" ORNL staff to conduct evaluation studies. The core issue is whether an "in-house" evaluation effort can be sufficiently objective or neutral in its conduct of research. This issue is part of a quite spirited debate over federal program evaluation that goes far beyond the DOE weatherization program.[159] A key assumption of the debate is that a separation of evaluation duties from program implementation duties would increase the "independence" of the evaluating organization. In response to criticism regarding ORNL, DOE engaged a competitive process to retain an outside contractor to conduct the second national evaluation (PY2007-PY2008) and the evaluation of the Recovery Act period (PY2009-PY2011). As noted previously, the competition led to the selection of the APPRISE contracting firm.

Debate over the Use of Evaluation Contractors

A recent debate between professors Charles Metcalf and David Reingold focused on the extent to which the use of "outside" contractors to conduct federal program evaluations can significantly improve research independence and objectivity.[160] In a separate study, Professor Jacob Klerman continues the dialogue, adding further perspectives to the debate. All three professors suggest ways to strengthen the independence of evaluations conducted with contractors.

Metcalf Argues: Use of Contractors Improves Evaluation Independence

Charles Metcalf speaks from experience as both a professor and an evaluation contractor. He says the central issue for evaluation independence is control over research processes and reporting. Metcalf acknowledges that outside pressures can affect a contractor's objectivity. In particular, he contends that client (agency/program staff) authority to accept or reject the evaluation findings and report—and to time its release—can give it significant control over evaluation content.[161] Metcalf argues that the independence and objectivity of contractor evaluations are promoted by four main conditions: maturity of the contractor organization, contractor reputation and longevity, formal agency-contractor agreements, and research transparency.[162] First, he claims that, over the past 40 years, the methods and practices of federal program evaluation have matured to the point where major evaluation contractor organizations have become generally objective and independent:

> Mathematica Policy Research (MPR), and others such as Abt Associates, MDRC, and RAND ... have established objectivity and independence as core values, and by and large have successfully defended these values in their conduct with funding organizations. Our track record establishes that government supported policy research can be and is independent, with proper safeguards.[163]

Second, Metcalf claims that the ability of a contractor organization to establish a reputation for objectivity enables it to enjoy longevity. In turn, this longevity helps the contractor to be independent from any single administration:

> If we define the client to be the current administration, the contractor and often the contract outlive the current client. For contract research institutions that regard themselves as doing policy research for a sequence of clients with differing policy agendas, the institutions' reputation for objectivity (as well as the quality of their work) is their primary currency in maintaining their longevity.[164]

Third, Metcalf argues that the agency's ability to influence content is largely constrained by the evaluation contract:

> For contract research, the researcher and his or her employing institution are collecting data and conducting research on an issue defined and paid for by a client. In this environment, two questions must be answered: (1) Who owns the research product? and (2) Who controls the

> content and the dissemination of the research product? These questions are typically defined by the contract, and the rights no longer lie intrinsically with the institution nor, by extension, with the researchers it employs to conduct the research.[165]

Further, he claims that there is an increased tendency for researchers and clients to agree explicitly, in advance, on the research methodology and the scope of reports. This process tends to protect the client (agency) from unexpected surprises, and protects the researcher (contractor) by allowing less "wiggle room" for the client to reinterpret results.[166] Fourth, Metcalf contends that research transparency is a key factor for contract evaluation independence. In this regard, he draws a parallel to academic research:

> [T]he basic guardian of both quality and objectivity in the conventional world of scholarly research is the required openness to the examination of others. If you don't adhere to this rule, your research is presumptively not credible.[167]

Metcalf argues that openness in the various phases of federal program evaluations is guaranteed by the government's competitive procurement process and by the Freedom of Information Act (FOIA). In procurement, the agency makes the request for proposals (RFP) publicly available. After the contract is awarded, the public can submit a FOIA request to obtain both the contract document (including the statement of the work) and the winning proposal:

> Regardless of internal contract restrictions, the contractor has the implicit nonexclusive right to request its own report under the FOIA, and to release it freely without restriction. Thus, while the government owns the research products it purchases through contract, the FOIA protects the researcher's right to disseminate his or her results—and, indeed, the right of anyone else to disseminate the results.[168]

Metcalf further claims that, in practice, the right of evaluation contractors to disseminate research results through conference participation and publication parallels the rights of researchers in academic institutions.[169] Also, in parallel to academic peer review, he notes that third parties often have a role in evaluations:

> First, we have already stressed the importance of the ultimate availability of research reports to the public, time delays notwithstanding.

Second, most large-scale studies engage advisory panels and other forms of peer review at both the design and report stages.[170]

Thus, he observes that third party involvement is an integral aspect of transparency. While acknowledging that threats to independence persist, Metcalf contends that the contracting environment has had a positive effect on evaluation. He concludes that the practice of using external evaluation contractors has shown an ability to enhance research credibility and independence.[171]

Reingold Counter-Argues: With Contractors, Threats to Independence Remain

Professor Reingold starts from the same assumptions as Professor Metcalf. Reingold says separation of the evaluation activity is assumed to establish evaluator independence which— along with quality research techniques— allows program evaluation to serve an accountability function:[172]

> [U]nlike other services frequently contracted out by governmental and nongovernmental organizations, the rationale for contracting out program evaluation rests largely in its ability to establish independence by creating a clear division of labor between those conducting evaluations and those responsible for managing programs being evaluated.[173]

He elaborates further that the process for selecting a contractor is vital to independence:

> [T]he perception of independence is typically established via the selection process used to identify and hire the person or organization responsible for designing and implementing the evaluation. For many organizations that purchase evaluation services, independence is established by selecting someone who is external to the organization or program being evaluated and who is free from financial self-interest or ideology that may bias the evaluation to produce a desired outcome, and where the evaluator has control (or ownership) over the evaluation design and its intellectual property (data and findings).[174]

Reingold argues that all of the “safeguards” to independence that Metcalf identified are overwhelmed by three main factors: (1) agency funding control gives it influence over evaluation content, (2) contractor organizations have financial self-interests in accommodating agency wishes, and (3) their mutual self-interests undermine the competitiveness of the selection process.

First, Reingold argues that an agency's control over evaluation funding allows it to retain a strong influence over evaluation content:

> [W]hen purchasing evaluation services via contracts, the purchasing agency has total control over all aspects of the services to be provided ... [thus] ... [i]t is inconsistent to argue that contracted program evaluation can be independent when the agency purchasing the evaluation owns (or controls) the research question and design, the method of data collection, the strategy of analysis, the data, the final report, and the rules governing the dissemination of results.[175]

Second, Reingold contends that contractor organizations have a financial self-interest in accommodating agency wishes:

> [E]valuation contractors view their relationships with those responsible for running programs and funding evaluations as clients where satisfaction with the deliverable will influence the likelihood of securing additional business in the form of more evaluation contracts.[176]

In other words, he argues, independence can be lost when experts that are expected to produce independent facts also have a substantial incentive to acquiesce to a client's interests. Third, Reingold claims that self-interests may begin to erode the contractor selection process, which underpins the independence sought by separating evaluation activity from program management:

> When program staff are allowed to strongly influence the selection of a firm to conduct an evaluation of their program, they frequently select firms that will "work with them" to produce a "helpful" evaluation. This is code for selecting a firm that will tailor and implement an evaluation design that will likely provide positive findings for the program under study.[177]

Over time, this threat to independence can deepen. Ultimately, he says, the complementary self-interests of agency and contractor may undermine the competitive foundation of the selection process:

> When a firm has established itself as a reliable partner that will cooperate with the demands of a particular client, the selection process has virtually eliminated any real competition that would allow for a truly independent evaluation.[178]

In sum, Reingold argues that threats to independence and objectivity can enter from both sides—a program's interest in self-protective control and a contractor's interest in preserving a business relationship. He concludes that:

> Unfortunately, there are signs this arrangement [the effort to achieve greater independence through external contracting] is not working.[179]

To address the problem, Reingold suggests two policy options to improve the independence of evaluation contracting. One option is to substitute grants for contracts, as the grants confer greater independence to the recipient. However, grants typically allow less oversight than contacts, which means that quality could suffer. Another option is to expand third party involvement. This option is described in the last section of the report.

Klerman Contributes to the Debate

Since the Metcalf-Reingold debate, there has continued to be a focus on means to reduce threats to evaluation independence. For example, Professor Jacob Klerman echoes some points from both Metcalf and Reingold in describing his view of the challenges to contractor evaluation independence. Klerman stresses two key threats to independence: the mutually reinforcing self-interests of funder and contractor and the self-interests of third parties. Also, he offers two ways to improve independence by modifying formal contract agreements. First, Klerman characterizes the main threat to independence as an "inherent tension" between funder and contractor.[180] Beyond stated goals for quality and objectivity,[181] he claims that both parties bring "unstated goals" involving pre-set policy opinions and organizational survival interests that can color an evaluation project's process and findings. He echoes Reingold's argument that self-interests pose a key threat to independence:

> [E]valuation is sometimes politics (or business) by other means. For the funder, showing that the current administration's programs "work" (and those of the previous administration did not "work") yields political benefits. Sometimes the concerns are quite direct. A negative evaluation result might cause shrinking of a program, loss of budget for the funder's organization, in the extreme even loss of jobs for the funder's lead evaluation staff. For the contractor, this is a business. The "right answer"—that is, the answer that the funder "wants to hear"—now is likely to lead to more evaluation business in the future; the "wrong answer" now is likely to lead to less evaluation business in the future.[182]

Second, Klerman proposes that when seeking a balance point between funder and contractor over evaluation content, one option is to involve a third party. However, "[e]ven nominating some third party leaves open the question of the third party's policy and organizational goals."[183] From his perspective, all three parties to an evaluation—funder, contractor, and third party—are propelled by self-interests that pose threats to independence. However, Klerman raises the question of whether there may be an "appropriate balance" of these interests that could actually reinforce independence.[184] In conclusion, Klerman—like Metcalf—claims that independence may be improved by augmenting, or otherwise improving, formal contract agreements. He describes one strategy for improvement:

> The challenge is to devise contractual mechanisms that assure satisfaction of the stated goal—a high-quality and independent evaluation. Ideally, a contract would give the funder enough authority to handle standard contractual issues while preventing either side from deviating from the stated goals to further policy or organizational goals.[185]

As a starting point for this strategy, Klerman suggests that a clear separation be established between an official evaluation report and a contractor's own publication of analysis from the underlying evaluation. He recommends that:

> The funder would retain almost unfettered rights to the official contract report. Those rights would include the right never to officially release the document but not the right to change the contractor's text while continuing to list the contractor and its individual staff members as authors. The contractor would retain clearly defined rights to publish any findings from the evaluation.[186]

Klerman's main strategy for improving independence encompasses 19 approaches to modify evaluation contracts, which are grouped into three categories: (1) approaches applied when a funder retains the right to specify the final report text, specify the report release, and control discussion of the process; (2) approaches that limit the funder's right to specify the final text, and (3) approaches that limit the funder's right to specify report release.[187] **Table 5** summarizes the main points of the Metcalf-Reingold debate, including the points contributed by Klerman.

The Limits of Independence: Is There an Optimum?

Agencies of other national governments are also concerned about program evaluation independence. An example is the United Kingdom's Department for International Development (DFID). An assessment of DFID's evaluation function addressed many of the same issues of control over the research process and findings that were the focus of the Metcalf-Reingold debate. The study observed that independence is important, a critical aspect of evaluation credibility and reliability:

> To be sure, independence on its own does not guarantee quality (relevant skills, sound methods, adequate resources and transparency are also required) but there is no necessary trade-off between independence, quality or credibility. Indeed, evaluation quality without independence does not assure credibility. Furthermore, in open and accountable working environments, evaluation independence induces credibility, protects the learning process and induces program managers and stakeholders to focus on results. Thus, evaluation independence, quality and credibility are complementary characteristics that together contribute to evaluation excellence.[188]

The DFID assessment of evaluation function suggests that the quest for independence—if taken too far—could lead to dysfunction. The report suggests that there may be limits to evaluation independence:

> Optimum independence is not maximum independence. Accurate and fair evaluations combine intellectual detachment with empathy and understanding. The ability to engage with diverse stakeholders and secure their trust while maintaining the integrity of the evaluation process is the acid test of evaluation professionalism. This is why diminishing returns set in when evaluation independence assumes extreme forms of disengagement and distance. Independence combined with disengagement increases information asymmetry, ruptures contacts with decision makers and restricts access to relevant sources of information. It leads to isolation, a lack of leverage over operational decision making and a chilling effect on learning. Thus, the basic challenge of evaluation governance design consists in sustaining full independence without incurring isolation.[189]

That finding appears to be a more elaborate expression of a key conclusion that Klerman makes in regard to achieving an "appropriate balance" of the three potential parties to a contract evaluation:

> [N]o single party—not the funder, not the contractor, and not some third parties—will always be without bias. Different approaches shift the relative power of the three groups. Furthermore, different approaches have different costs—in dollars, in the formality of the funder–contractor relationship, and in time to complete the study. The appropriate balance will vary with the particular situation, and, in particular, with the prevalence of problematic behaviors on all sides—funder, contractor, and third parties.[190]

As noted previously, Klerman proposes several means to modify the client-contractor relationship to address the inherent tension and reduce—but not eliminate—the threats to objectivity and independence.[191]

Implications of the Independence Debate for DOE Evaluations

DOE Addresses Threats to Independence

DOE is aware of the concern about contractor independence for impact evaluations. It has published guidance for program staff that addresses the issue:

> Evaluation should be conducted by outside independent experts. This means that, even if staff commission a study (fund an evaluation contractor) that contractor should have some degree of independence from the program office that is being evaluated. Also, the contractor should have no real or perceived conflict of interest. Although the program staff person may work with the contractor during the consultation phase to clarify information needs and discuss potential evaluation criteria and questions, the staff should establish some line of separation from the contractor for much of the remainder of the evaluation study—i.e., put up a firewall after the initial consultation period is concluded.[192]

Also, in order to address remaining concerns about outside contractor independence, DOE guidance further recommends the establishment of a third party group of experts to critique the contractor's work:

> It is further highly recommended that a panel of external evaluation experts who are not part of the contractor's team be assembled to serve as reviewers of the contractor's work. This would include the Evaluation Plan, and the draft and final reports. Having the work of the contractor's team itself evaluated helps ensure the evaluation methodology is sound

and the study is carried out efficiently. It also sets up a second firewall that raises the credibility of the study to even a higher level (important for those who remain skeptical of evaluation studies commissioned by the program being evaluated).[193]

Further Considerations for DOE's Active Evaluations

DOE has addressed the first point of the independence debate. It separated the evaluation activity from the program activity by choosing outside contractors to perform a strategic assessment and two major evaluation studies. Additional comments and questions may focus on the other three points of the independence debate:

- Contractor Selection. All three debaters stressed that the selection process is critical to evaluation independence. Was the selection of Deloitte and APPRISE competitive? What objective indicators should be used to gauge the competitiveness of the selection process? Is there an optimum degree of selection process competitiveness?
- Transparency. Metcalf argued that transparency helps guarantee evaluation independence. Will this dimension be fulfilled by the public release of the three final reports? Otherwise, what objective indicators should be used to gauge transparency for each evaluation? Is there an optimum degree of transparency?
- Third Party Involvement. Klerman described the possibility of an "appropriate balance" that included a third party. Related to the ongoing evaluation of the Recovery Act period (2009-2011), the DOE IG issued a progress report and GAO has contributed a major performance audit. Is that sufficient third party involvement? Otherwise, what objective indicators should be used to gauge the sufficiency of third party involvement for each evaluation? Is there an optimum degree of third party involvement?

Some Policy Options That May Enhance Independence

The debate between Metcalf and Reingold produced some other ideas for establishing independence of the evaluation function.[194] One notable strategy suggests using other (third party) organizations to play a more central role in managing the program evaluation process. Reingold suggests:

> Ideally, Congress, the GAO, and perhaps the National Academy of Sciences [National Research Council] could play a more active role in screening and selecting evaluation plans and firms.[195]

Three variations on this strategy, in increasing order of dependence on the outside agency are:

(1) Have the DOE Office of the Inspector General (DOE IG) or GAO or Congressional Budget Office (CBO) ensure that the third party contractor evaluation work is performed in an independent manner.
(2) Move the contractor (evaluator) selection process from the agency to the Office of the Inspector General, GAO, CBO, or the National Research Council (NRC).
(3) Transfer full responsibility for conduct of the evaluation to the DOE IG, GAO, CBO, or NRC.

Table 5. Debate: Does Evaluation Contracting Promote Independence? (Metcalf – Reingold Arguments)

Issue Premise: Pressures exist that threaten the independence and objectivity of evaluation studies. In order to fulfill an accountability function, evaluation studies must be conducted independently of agency and program influences.		
Dimension/ Debate Point	**Contracting promotes evaluation independence. (Professor Metcalf)**	**Contracting does not guarantee evaluation independence. (Professor Reingold)**
separate evaluation activity	The division of labor promotes independence.	Contractors have a self-interest in preserving their business/ organization, which makes them "bend" to program/agency influence.
select contractor competitively	Competitive procurement of contractors promotes independence.	The contractor selection process determines independence. Funding control gives the program/agency influence over contractor selection. Also, the competitiveness of the selection process may be undermined by organizational self-interests on both sides.
make evaluation transparent	Research process transparency promotes independence. Prior agreements over research design and methods helps ensure independence and objectivity. Responses to RFPs are publicly available. FOIA allows contractors and others to make research results publicly available.	Funding control gives influence over research design, data collection, analysis, reporting, and dissemination.
involve third party	Third party/peer review activity promotes independence.	Third parties also have organizational self-interests.

Source: Richard P. Nathan, Journal of Policy Analysis and Managerment, (Point/Counterpoint) *Can Government-Sponsored Evaluations Be Independent?* v. 27, no. 4, pp. 926-944, 2008. Also see Klerman, Jacob Alex, *Contracting for Independent Evaluation: Approaches to an Inherent Tension. Evaluation Review*, July 20, 2010, pp. 299-333.

Notes: Although the original debate was framed by a 2008 point-counterpoint dialogue between Metcalf and Reingold; in 2010, Klerman made substantial contributions to the debate, which are incorporated into the table.

APPENDIX A. DOE WEATHERIZATION PROGRAM: HISTORICAL DATA

Table A-1. DOE Weatherization Funding, Units Weatherized, and Benefit-Cost Ratio ($ millions)

Fiscal Year	Request $Current	Approp. $Current	Deflator $2005	Deflator $2010	Request $ 2010	Approp. $2010	Annual Units	Cumulative Units	Benefit-Cost Ratio
1977	0.0	27.5	0.3750	0.3362	0.0	81.6	1,622	1,622	
1978	54.1	65.0	0.4003	0.3589	150.4	180.7	6,742	8,364	
1979	198.5	199.0	0.4325	0.3878	510.7	511.8	15,387	23,751	
1980	199.0	199.0	0.4707	0.4220	470.4	470.3	232,751	256,502	
1981	175.0	175.0	0.5171	0.4636	376.6	376.5	352,906	609,408	
1982	0.0	144.0	0.5525	0.4954	0.0	290.0	122,992	732,400	
1983	0.0	245.0	0.5768	0.5172	0.0	472.6	156,629	889,029	
1984	0.0	190.0	0.5981	0.5363	0.0	353.5	209,261	1,098,290	
1985	190.0	191.1	0.6175	0.5537	342.4	344.4	163,860	1,262,150	
1986	152.9	182.1	0.6318	0.5665	269.3	320.7	149,047	1,411,197	
1987	0.0	161.3	0.6486	0.5815	0.0	276.7	105,440	1,516,637	
1988	0.0	161.3	0.6694	0.6002	0.0	268.1	105,465	1,622,102	
1989	0.0	161.3	0.6954	0.6235	0.0	258.1	85,115	1,707,217	1.06
1990	0.0	162.0	0.7210	0.6465	0.0	250.0	84,441	1,791,658	
1991	15.0	198.9	0.7483	0.6709	22.3	295.8	105,769	1,897,427	
1992	24.0	194.0	0.7678	0.6884	34.8	281.1	99,587	1,997,014	
1993	80.0	185.4	0.7848	0.7037	113.4	262.9	103,394	2,100,408	
1994	239.4	214.8	0.8014	0.7186	332.4	298.2	114,904	2,215,312	
1995	249.8	226.4	0.8184	0.7338	339.6	307.8	102,981	2,318,293	

Table A-1. (Continued)

Fiscal Year	Request $Current	Approp. $Current	Deflator $2005	Deflator $2010	Request $ 2010	Approp. $2010	Annual Units	Cumulative Units	Benefit-Cost Ratio
1996	229.0	111.7	0.8342	0.7480	305.5	149.0	76,393	2,394,686	1.79
1997	155.5	120.8	0.8495	0.7617	203.7	158.3	71,597	2,466,283	
1998	154.1	124.8	0.8603	0.7714	199.3	161.4	68,470	2,534,753	
1999	154.1	133.0	0.8717	0.7816	196.7	169.8	71,984	2,606,737	1.51
2000	154.0	135.0	0.8889	0.7970	192.8	169.0	74,316	2,681,053	
2001	154.0	153.0	0.9099	0.8158	188.3	187.1	77,709	2,758,762	
2002	273.0	230.0	0.9249	0.8293	328.4	276.7	104,860	2,863,622	1.30
2003	277.1	223.5	0.9442	0.8466	326.6	263.4	100,428	2,964,050	
2004	288.0	227.2	0.9684	0.8683	330.9	261.0	99,593	3,063,643	
2005	291.2	228.2	1.0000	0.8966	324.0	253.9	97,500	3,161,143	1.34
2006	230.0	242.6	1.0342	0.9273	247.5	261.0	104,283	3,265,426	1.54
2007	164.2	204.6	1.0643	0.9543	171.5	213.6	89,772	3,355,198	1.54
2008	144.0	227.2	1.0890	0.9764	147.0	232.0	6,116	3,361,314	1.65
2009	0.0	5,427.5	1.1054	0.9911	0.0	5,468.8			1.67
2010	220.0	210.0	1.1153	1.0000	220.0	210.0			1.80
2011	300.0	171.0			296.1	168.8			
2012	320.0				311.5				
Total		11,453.0				14,504.5			
Subtotal, 1977-2008		5,644.5				8,657.0			

Source: DOE Weatherization Budget History; ExpectMore PART Results; and various evaluation studies.

Appendix B. DOE Weatherization Funding Chart

Figure B-1, below, shows the entire history of annual requests and appropriations for the DOE Weatherization Assistance Program (WAP). Requested amounts for FY1977, FY1978, and FY1979 were not available. A final FY2012 appropriation figure had not yet been determined at the time this report was prepared.

($ millions, FY2010 constant dollars)

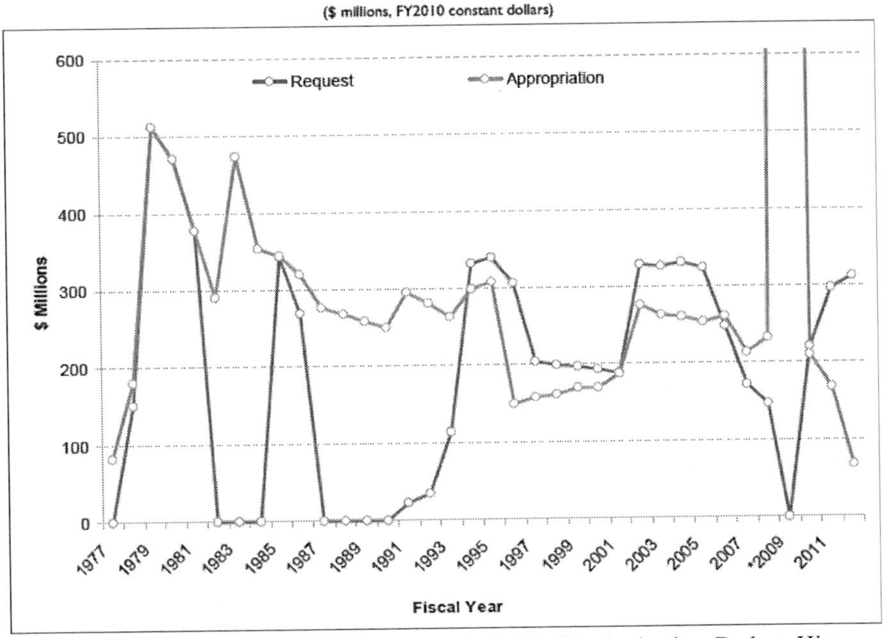

Source: DOE budget requests (various years); DOE Weatherization Budget History; and ExpectMore PART Results.

Notes: For FY2009, WAP received an appropriation of $427.5 million ($ current) plus $5 billion ($ current) from the Recovery Act (P.L. 111-5). The total amount is toolarge to be shown on this chart, but the funding amounts appear in Appendix A, **Table A-1**.

Figure B-1. DOE Weatherization Program: A Review of Funding and Cost-Effectiveness Studies.

End Notes

[1] More detailed background is available in DOE's *Weatherization Assistance Program Overview* at http://www.waptac.org/data/files/website_docs/briefing_book/wap_programoverview_final.pdf.

[2] Title 42 of the U.S. Code, Chapter 81, Subchapter III, Part A, 6861.

[3] DOE, *History of the Weatherization Assistance Program*. http://www1.eere.energy.gov/wip/wap_history.html

[4] In 1977, Congress first appropriated funds to help low-income households pay energy bills. In 1981, P.L. 97-35 gave this activity statutory authority as the LIHEAP program.

[5] For further information on LIHEAP, see CRS Report RL31865, *The Low Income Home Energy Assistance Program (LIHEAP): Program and Funding*, by Libby Perl

[6] DOE, *History of the Weatherization Assistance Program*.

[7] Ibid.

[8] ORNL, *Weatherization Assistance Program Technical Memorandum Background Data and Statistics* (ORNL/TM-2010/66, prepared by Joel Eisenberg) March 2010, p. 3. The Program now defines eligibility as household income at or below 200% of the poverty income guidelines or of the HHS LIHEAP eligibility guidelines. Of the 111 million households nationwide, about 38.6 million are eligible for LIHEAP, and about 16.6 million have an income below the poverty level. The geographic distribution is roughly 21% in the Northeast, 24% in the Midwest, 36% in the South, and 19% in the West. http://weatherization.ornl.gov/pdfs/ORNL_TM-2010-66.pdf

[9] Ibid, p. 5. The energy burden was defined as average residential energy expense divided by average income.

[10] DOE, *Weatherization Assistance Program Technical Memorandum Background Data and Statistics*, March 2010, p. 5. RECS stands for the Residential Energy Consumption Survey prepared by DOE's Energy Information Administration (EIA). http://weatherization.ornl.gov/pdfs/ORNL_TM-2010-66.pdf

[11] The Energy Independence and Security Act of 2007 (P.L. 110-140, §411c) added Puerto Rico and the territories of the U.S. to the definition of "State" for the purpose of funding allocations. Beginning with Program Year 2009, the territories of American Samoa, Guam, Commonwealth of the Northern Mariana Islands, Commonwealth of Puerto Rico and the U.S. Virgin Islands were added to the program.

[12] Administrative rules, eligibility standards, the types of aid, and benefit levels are primarily decided at the state level. Eligibility is automatically given to applicants receiving Temporary Assistance to Needy Families (TANF) or Supplemental Security Income (SSI). Also, if a state elects, program eligibility can extended to a household that meets (LIHEAP) eligibility criteria.

[13] Most of the grantees are state-designated community action agencies, which administer multiple types of social service grants for low-income persons. No more than 10% of grant funds allocated to states may be used for administration.

[14] DOE, *Weatherization Assistance Program Funding,* p. II-12.

[15] The sum of the base allocations for all states totals $171,258,000. The total formula allocations equal the total program allocations minus the base allocations.

[16] DOE, *FY2011 Budget Request,* vol. 3, p. 440.

[17] DOE, *Weatherization Assistance Program Funding,* p. II-12. The approximation is necessary due to the lack of state-specific data on residential energy expenditures by low-income households.

[18] The interim final rule on the revised allocation formula, which was issued on June 5, 1995, set the minimum threshold at $209.7 million.

[19] DOE, *Weatherization Assistance Program Funding,* p. II-12. The rules state that if the post-TTA allocation falls below the threshold of $209.7 million set under P.L. 103-332, then

each state's program allocation shall be reduced by the same percentage relative to its allocation under P.L. 103-332. For example, if total program allocations for a given year were 10% below the amount under P.L. 103-332, then each state's program allocation would be 10% less than under P.L. 103-332.

[20] Unless otherwise noted, all funding figures refer to constant FY2010 dollars.

[21] The program funding from 1977 through 2008 can also be found at http://www.waptac.org/data/files/website_docs/briefing_book/2_programfunding_final.pdf.

[22] In constant 2010 dollars.

[23] DOE, *FY1983 Budget Request* (v. 3, Energy Conservation Grant Programs), February 1982.

[24] The White House, *National Energy Policy,* Report of the National Energy Policy Development Group, May 2001, p. xv, http://www.whitehouse.gov/energy/National-Energy-Policy.pdf.

[25] The White House, *National Energy Policy*, p. 2-3.

[26] The report also called for increases to funding for the Low Income Home Energy Assistance Program (LIHEAP) at the Department of Health and Human Services.

[27] The White House, *National Energy Policy*, p. 2-12.

[28] The assessment was obtained from use of the Program Assessment Rating Tool (PART), which was developed by the Office of Management and Budget to provide a standardized way to assess federal programs.

[29] DOE, *FY 2009 Congressional Budget Request*, vol. 3, p. 463.

[30] EERE's technology programs include Vehicle Technologies, Building Technologies, and Industrial Technologies.

[31] DOE, *FY 2009 Congressional Budget Request*, vol. 3, p. 44.

[32] DOE, Office of the Inspector General, *Special Report: Progress in Implementing the Department of Energy's Weatherization Assistance Program Under the American Recovery and Reinvestment Act* (OAS-RA-10-04), February 2010, p. 1. This was a major increase over the $450 million appropriated for the program in FY2009. http://www.ig.energy.gov/documents/OAS-RA-10-04.pdf

[33] DOE, *History of the Weatherization Assistance Program.*

[34] DOE IG, *Special Report on the Weatherization Program*, p. 5.

[35] Ibid.

[36] Davis-Bacon is the common name applied to legislation enacted in the 1930s that requires all federal construction projects to pay prevailing wages. For more background on the Davis Bacon Act, see CRS Report R40663, *The Davis-Bacon Act and Changes in Prevailing Wage Rates, 2000 to 2008*, by Gerald Mayer.

[37] DOE IG, *Special Report on the Weatherization Program.*

[38] Ibid. p. 4.

[39] DOE Inspector General, *Special Report,* p. 2.

[40] DOE Inspector General, *Special Report*, p. 5.

[41] For more on the definitions—and complementary aspects—of performance assessment and program evaluation studies, see: GAO, *Performance Measurement and Evaluation: Definitions and Relationships*, May 2011, http://www.gao.gov/assets/80/77277.pdf; and OMB, *Performance Measurement Challenges and Strategies*, June 18, 2003, http://www.whitehouse.gov/sites/default/files/omb/part/challenges_strategies.pdf.

[42] OMB's GPRA guidance is available on its website at http://www.whitehouse.gov/omb/mgmt-gpra/gplaw2m#h1.

[43] OMB, *Circular No. A-11,* July 21, 2010, Section 200.2, http://www.whitehouse.gov/sites/default/files/omb/assets/a11_current_year/a_11_2010.pdf.

[44] OMB, *Circular No. A-11,* Section 200.2.

[45] DOE, *FY2009 Annual Performance Report.* For the Weatherization Program, this report only cites one performance measure, the number of homes weatherized in FY2009, compared with the number weatherized in FY2006 through FY2008. p. 87. http://www.cfo.doe.gov/cf1-2/2009APR.pdf

[46] DOE's *FY2006 Performance and Accountability Report* appears to be the most recent one available. It reports a few selected findings about the Weatherization Program from OMB's PART assessment on pp. 11, 19, and 102. http://www.cfo.doe.gov/progliaison/2006 finalpar2.pdf

[47] DOE, *Strategic Plan (2006),* p. 27. http://www.cfo.doe.gov/strategicplan/docs/2006Strategic Plan.pdf

[48] OMB, *Circular No. A-11,* Section 200.4.

[49] Ibid.

[50] Government Accountability Office (GAO), *Government Auditing Standards: 2003 Revision,* June 2003 (Section 2.09). GAO provides a more detailed definition, as follows: Performance audits entail an objective and systematic examination of evidence to provide an independent assessment of the performance and management of a program against objective criteria as well as assessments that provide a prospective focus or that synthesize information on best practices or crosscutting issues. Performance audits provide information to improve program operations and facilitate decision making by parties with responsibility to oversee or initiate corrective action, and improve public accountability. Performance audits encompass a wide variety of objectives, including objectives related to assessing program effectiveness and results; economy and efficiency; internal control; compliance with legal or other requirements; and objectives related to providing prospective analyses, guidance, or summary information. Performance audits may entail a broad or narrow scope of work and apply a variety of methodologies; involve various levels of analysis, research, or evaluation; generally provide findings, conclusions, and recommendations; and result in the issuance of a report. http://www.gao.gov/govaud/yb2003.pdf#Page=27

[51] DOE generally defines a program year as the 12-month period starting on July 1 of the year after DOE receives the appropriation from Congress.

[52] DOE, Office of the Inspector General, *Audit of the Weatherization Assistance Program.* (Audit Report No.: OAS-L-03-15) April 18, 2003, p. 1. (Hereinafter referred to as DOE IG.)

[53] GAO, *Government Auditing Standards* (July 2007 Revision). http://www.gao.gov/govaud/ govaudhtml/index.html

[54] DOE IG, p. 1.

[55] Ibid.

[56] Ibid. p. 2.

[57] Ibid. p. 2.

[58] Office of Management and Budget (OMB). *Detailed Information on the Weatherization Assistance Assessment.* Weatherization Assistance. Program Assessment Summary. 2003. (Hereinafter referred to as "OMB, *Detailed Assessment.*") http://www.whitehouse.gov/omb/ expectmore/summary/10000128.2003.html

[59] This was an initiative of the George H.W. Bush Administration.

[60] DOE, *Performance and Accountability Report, Fiscal Year 2006.* (DOE/CF-0012) p. 11. http:// www.cfo.doe.gov/progliaison/2006finalpar2.pdf

[61] OMB, *What Constitutes Strong Evidence of a Program's Effectiveness?* 2004. http://www. whitehouse.gov/sites/default/files/omb/part/2004_program_eval.pdf

[62] DOE, *Performance Report Highlights FY2006,* p. 11.

[63] OMB, *Detailed Assessment.*

[64] Weatherization Assistance, Program Assessment Summary, Tab for "RATING: What This Rating Means." http://www.whitehouse.gov/omb/expectmore/rating.html

[65] OMB described each evaluation performed by ORNL as an "internal program assessment," because it was not conducted by an independent third party. The issue of independence is treated in more detail later, in another section.

[66] These issues were raised in the 2003 IG performance audit, which is reviewed in the previous section.

[67] OMB, *Detailed Assessment.*

[68] OMB cited figures that exclude values for non-energy benefits.

[69] The value of energy was discounted over future years at a rate of 3.2%.
[70] The program costs excluded some management costs.
[71] The price data were obtained from DOE's Energy Information Administration (EIA).
[72] The range for the benefit-cost ratio was tested for the 90% confidence range.
[73] ORNL benefit-cost estimates are discussed in the next section and are shown in **Appendix A**.
[74] OMB, *Weatherization Assessment*, Question # 2.6.
[75] Peer reviewers of ORNL weatherization reports are generally employees of ORNL or DOE. Ibid.
[76] The cited evidence was: State-level Evaluation of WAP 1990-1996: A Meta Evaluation of 17 States Evaluations -1997 (ORNL/CON-435); Metaevaluation of National Weatherization Assistance program Based on State Studies, 1996-1998 (ORNL/CON-467); Metaevaluation of National Weatherization Assistance Program Based on State Studies, 1993-2002 (ORNL/CO-488); Nonenergy Benefits from the Weatherization Assistance Program: A Summary of Findings from the Recent Literature, 2002 (ORNL/CON-484).
[77] GAO, *Government Auditing Standards* (Definition of Performance Audits in Chapter 1: Use and Application of GAGAS), July 2007 Revision (GAO-07-731G. http://www.gao.gov/govaud/govaudhtml/d07731g-3.html
[78] Ibid.
[79] The American Reinvestment and Recovery Act of 2009 (P.L. 111-5), Sections 901 and 902, H.Rept. 111-16, p. 77-78 and 463.
[80] GAO reported that the audit was conducted from September 18 through December 4, 2009. The review covered about two-thirds of federal assistance funds and about 65% of the national population. GAO, *Recovery Act: Status of States' and Localities' Use of Funds and Efforts to Ensure Accountability*, (GAO-10-231) December 2009, http://www.recovery.gov/accountability/documents/d10231.pdf.
[81] Ibid. p. 3-4.
[82] Ibid. p. 91-93.
[83] For this phase, GAO collected updated state interview data from nine states and the District of Columbia, interviewed 31 local service providers, and conducted several site visits. GAO, *Recovery Act: States' and Localities' Uses of Funds and Actions Needed to Address Implementation Challenges and Bolster Accountability* (GAO-10-604)May 2010, http://www.gao.gov/new.items/d10604.pdf
[84] GAO recommended that DOE clarify program guidance to address issues involving income eligibility, training standards, state monitoring, multi-family units, local agency internal controls, and methods for assessing costs and cost-effectiveness. Ibid. p. 103-125.
[85] Personal communication with Arvin Wu at GAO. January 2011.
[86] GAO, *Recovery Act: Progress and Challenges in Spending Weatherization Funds*, (GAO-12-195) December 2011, http://www.gao.gov/products/GAO-12-195.
[87] The 58 recipients include the 50 states, seven territories, and District of Columbia.
[88] Ibid. p. 2-4.
[89] Ibid. p. 9.
[90] Ibid. p. 19-22.
[91] Ibid. p. 17 and 31-34. GAO does not make an independent estimate of cost-effectiveness. Instead, it refers to an ORNL Technical Memorandum issued March 2010 (reviewed in a later section of this report, on p. 26) that provided a preliminary estimate of cost-effectiveness, including a benefit-cost ratio of 1.8 for energy savings, rising to 2.5 with non-energy benefits included.
[92] Ibid. p. 10 and 14.
[93] Ibid. p. 35-36.
[94] Many of ORNL's evaluation reports on the Weatherization Program are available at http://weatherization.ornl.gov/Publications.htm.
[95] Subsequently, ORNL conducted four "metaevaluations" of the Program's energy savings, which used studies conducted by individual states between the years 1990–1996, 1996–

1998, 1993–2002, and 1993-2005. For more details about those studies, see: ORNL, *National Evaluation of the Weatherization Assistance Program: Preliminary Evaluation Plan for Program Year 2006,* February 2007, p. 1. http://weatherization.ornl.gov/pdf/ Prelim%20Eval%20Plan-Feb%202007.pdf

[96] ORNL managed the five-part study, which was based mainly on data from PY1989, supplemented with data from 1991 and 1992.The 400-page report is available at http://weatherization.ornl.gov/pdfs/ORNL_CON-326.pdf.

[97] Home Energy. *Evaluating DOE's Weatherization Assistance Program.* July/August 2010. At that time, the program weatherized about 250,000 dwellings per year through approximately 1,110 local weatherization agencies. The evaluation consisted of five parts: a network study, a resources and population study, a multifamily study, a single-family study, and a fuel oil study. The average savings for all single-family and small multifamily dwellings in 1989was estimated to be 17.6 million Btu per year, 18.2% of the energy used for space heating and 13.5% of total energy use. It was estimated that over a 20-year lifetime, the program would save the equivalent of 12 million barrels of oil. http://www.homeenergy. org/article_full.php?id=725

[98] ORNL, *Dear Colleague Letter from Marilyn A. Brown* [To accompany National Weatherization Evaluation Report], October 1993.

[99] More recent DOE reports appear to suggest that the benefit-cost figure was later revised to a value of 1.06.

[100] A metaevaluation involves synthesizing results from a number of individual studies of state weatherization efforts in order to estimate total national impacts.

[101] ORNL, *State-Level Evaluations of the Weatherization Assistance Program in 1990-1996: A Metaevaluation that Estimates National Savings,* (ORNL/CON-435) January 1997. [http://weatherization.ornl.gov/pdfs/ORNL_CON-435.pdf].

[102] The 15 states included 9 with published results—Colorado, Indiana, Iowa, New York, North Carolina, North Dakota, Ohio (2 studies), Texas, and Vermont (2 studies)—and 6 with unpublished results, including Delaware, Kansas, Minnesota, Nebraska, Wisconsin, and Wyoming.

[103] Ibid. p. xii.

[104] Ibid. p. xi.

[105] Ibid. p. 11. Three types of evidence supported the finding that savings were increasing: a literature review, within-state comparisons of savings over time, and regression results.

[106] Ibid. p. 16.

[107] Ibid. p. 17.

[108] Ibid. p. 18.

[109] ORNL, *Metaevaluation of National Weatherization Assistance Program Based on State Studies, 1996-1998,* (ORNL/CON-467) 1999. ORNL's review included studies of Colorado, Delaware, District of Columbia, Indiana, Iowa (2), Minnesota (2), Ohio, and Vermont. Of the 10 state studies, 3 used control groups and 7 did not. pp. 4-5.

[110] ORNL, *Metaevaluation 1996-1998*, p. x. The key variable(s) associated with energy savings were identified by running a regression analysis using energy savings as the dependent variable and a number of potentially related factors as independent variables. Estimates of average household energy savings nationwide were accomplished by taking the regression equation frorn the model with the best predictive ability and inserting the average national values for the independent variable(s).

[111] Ibid. p. 14.

[112] ORNL, *Metaevaluation 1996-1998*, p. 1.

[113] Personal communication with Robert Adams, Supervisor of the DOE Weatherization Program, January 30, 2011. The 2005 Metaevaluation is at [http://weatherization.ornl.gov/pdfs/ ORNL_CON-493.pdf].

[114] ORNL, *Estimating the National Effects of the U.S. Department of Energy's Weatherization Assistance Program With State-Level Data: A Metaevaluation Using Studies From 1993 to*

2005, (ORNL/CON-493), September 2005. http://weatherization.ornl.gov/pdfs/ORNL_CON-493.pdf

[115] The 19 states were: Colorado, Delaware, District of Columbia, Georgia, Illinois, Indiana, Iowa, Kansas, Minnesota, Nebraska, New York, North Carolina, Ohio, Texas, Vermont, Washington, West Virginia, Wisconsin, and Wyoming.

[116] ORNL, 2005, p. xi and 4.

[117] Ibid. p. 3. This metaevaluation included new data on measured energy savings from six states: Delaware, District of Columbia, Iowa, Illinois, Texas, and Wisconsin.

[118] Ibid. p. 3-4.

[119] Ibid. p. xi.

[120] Ibid. p. xii.

[121] Within the sample of 19 states, studies in 17 states addressed savings in dwellings heated by natural gas.

[122] Ibid. p. 15.

[123] Ibid. p. 5.

[124] Ibid. p. xiv.

[125] ORNL, *Weatherization Assistance Program Technical Memorandum Background Data and Statistics,* (ORNL/TM-2010/66), Prepared by Joel Eisenberg, March 2010, p. 1. http://weatherization.ornl.gov/pdfs/ORNL_TM-2010-66.pdf

[126] NEAT was developed by ORNL. It is part of "The Weatherization Assistant," a family of computer audit software programs that DOE designed to help state and local agencies implement the program. The Weatherization Assistant serves as the umbrella program for NEAT and for DOE's Manufactured Home Energy Audit (MHEA). *The Weatherization Assistant Features*, http://weatherization.ornl.gov/assistant_features.shtml.

[127] Ibid. p. 7.

[128] ORNL, *Technical Memorandum*, p. 7.

[129] ORNL, *Metaevaluation 2005.*

[130] ORNL, *Technical Memorandum*, p. 7.

[131] DOE, WAPTAC, *Simple Savings to Investment Ration (SIR) Comparison.* Actual SIR calculations supported by NEAT, MHEA, and other approved audit tools account for the present value (PV) of money and fuel escalation rates over the lifetime of the measures to arrive at more accurate savings numbers. DOE describes the SIR in a worksheet available online at http://www.google.com/search?rlz=1C1CHFX_enUS389US389&aq=f&sourceid=chrome&ie=UTF-8&q=savings+investment+ratio.

[132] ORNL, *Technical Memorandum*, p. 7.

[133] Ibid. p. 9

[134] Ibid. p. 9.

[135] Ibid. p.13.

[136] Ibid. p. 13.

[137] Deloitte Touche Tohmatsu International is a major, international accounting and auditing firm. Deloitte LLP is the U.S. subsidiary of the international firm.

[138] DOE, *Weatherization Program Notice 09-1*, November 17, 2008.

[139] DOE, *Weatherization Program Briefing Book.* National Evaluation. 2006. http://www.nyswda.org/DirManual/Documents/06DOEBriefingBook/Issues.pdf

[140] Ibid.

[141] ORNL, *National Evaluation of the Weatherization Assistance Program*, 2007. p. 1. The report says that DOE undertook the new evaluation because the program is now "vastly different" from the PY1989 program that was examined in the 1993 national evaluation.

[142] Ibid.

[143] Home Energy, *Evaluating DOE's Weatherization Assistance Program,* June 30, 2010. APPRISE is a nonprofit energy research company. Key team members include the Energy Center of Wisconsin, Michael Blasnik & Associates, and Dalhoff Associates, LLC.

[144] As of early 2011, there was an indication that the study might produce some early results for PY2007 and PY2008. As of December 2011, such results had not been published. There was also an early 2011 expectation that the study would include some PY 2009 and PY 2010 results, which would become available in following years. Personal communication with Robert Adams, DOE Weatheriztion Program Office, January 2011.

[145] ORNL. *Evaluation of the National Weatherization Assistance Program during Program Years 2009-2011 (American Reinvestment and Recovery Act Period),* ORNL/TM-2011/87, May 20, 2011 (draft date), pp. 1 and 123.

[146] Home Energy, *Evaluating DOE's Weatherization Assistance Program*. DOE reports that it is difficult to measure the impact of energy efficiency programs on homes that use bulk fuels, such as propane and fuel oil. This is because residents use different suppliers and sometimes use additional fuels, and because they do not always fill up the tank completely. To meet this challenge, the evaluation team will install submeters to measure energy savings in homes heated with bulk fuels.

[147] Ibid.

[148] Ibid.

[149] Ibid.

[150] ORNL is developing an evaluation plan for the Recovery Act funding period that it is coordinating with APPRISE.

[151] ORNL, *Technical Memorandum*, p. 1. Partial results were expected to emerge as early as late 2011.

[152] GAO, *Performance Audit for Recovery Act Spending on DOE Weatherization Program*, p. 34.

[153] Home Energy, *Evaluating DOE's Weatherization Assistance Program.*

[154] ORNL, *Evaluation of the National Weatherization Assistance Program during Program Years 2009-2011 (American Reinvestment and Recovery Act Period)*, ORNL/TM-2011/87, May 20, 2011 (draft date), pp.1 and 123.

[155] ORNL notes that Program Year (PY) 2009 started in April 2009 and PY2011 ends in June 2012. Ibid. pp.122-123.

[156] The weatherization program was not subject to Davis-Bacon rules before the Recovery Act.

[157] For example, did the Recovery Act funding change the use of requests for proposals (RFPs) versus bids?

[158] Home Energy, *Evaluating DOE's Weatherization Assistance Program.*

[159] The current debate over the use of outside evaluation contractors is part of a long-term historical debate over the relative merits of having the evaluation activity conducted by an organization located inside (internal evaluator)—or outside (external evaluator)—of the institution that operates the program. The internal versus external issue, and the debate over evaluation contractors, continues to be a concern for many federal programs.

[160] Richard P. Nathan, Journal of Policy Analysis and Management, (Point/Counterpoint) *Can Government-Sponsored Evaluations Be Independent?* v. 27, no. 4, pp. 926-944, 2008.

[161] Metcalf says this "ownership power" is the client's primary way to influence content. Ibid. p. 930.

[162] Metcalf, Charles E. *Threats to Independence and Objectivity of Government-Sponsored Evaluation and Policy Research.* Journal of Policy Analysis and Management. vol. 27, 2008. p. 927-934.

[163] RAND and MDRC are both non-profit, non-partisan policy research organizations. The RAND acronym was originally formed by a contraction of the term "research and development." The MDRC acronym was originally formed from the term, Manpower Demonstration Research Corporation. Ibid, p. 927.

[164] Ibid, p. 932.

[165] Ibid, p. 928.

[166] Also, he says this environment has promoted greater use of experimental research methods, which leave less room for subjective judgment than non-experimental methods. Ibid, p. 932.

[167] Ibid, p. 928.

[168] Since federal contracts currently tend to restrict automatic dissemination rights before the contract ends, Metcalf finds FOIA-guaranteed access very important. Ibid, p. 930.

[169] Ibid, p. 933.

[170] The use of advisory panels and related mechanisms can contribute a form of "third party" balance. Ibid, p. 931.

[171] Ibid, p. 933-934.

[172] Ibid, p. 934.

[173] Ibid, p. 935.

[174] Ibid, p. 935.

[175] Journal of Policy Analysis and Management. *David A Reingold's Response to Charles E. Metcalf*, Fall 2008, pp. 937 and 942. Reingold notes that program evaluation is not the only realm of scientific analysis that is subject to the problem of contracting bias.

[176] Ibid, p. 937.

[177] Ibid, p. 939.

[178] Ibid, p. 937.

[179] Ibid, p. 935.

[180] He points out that this unconventional nature is similar to that for certain other products. Auditing is cited as the leading example, where the tension arises because management is the direct client, but the audit will also be used by the broader investing community. Reputation and financial liability are noted as key incentives for auditing firms to show objectivity and independence. He cites the case of Enron and Arthur Andersen as an example where organization interests overwhelmed those incentives. Other such ''gatekeepers'' are observed, including lawyers in merger negotiations, bond raters, and securities analysts. Klerman, Jacob Alex, *Contracting for Independent Evaluation: Approaches to an Inherent Tension*, Evaluation Review, July 20, 2010, pp. 299-333.

[181] Klerman defines objectivity in the methodological sense, to mean "conclusions following from the analysis, no more and no less." Ibid. p. 329.

[182] Ibid. p. 302.

[183] Ibid. p. 304.

[184] Ibid. p. 310.

[185] Ibid, p. 304.

[186] Ibid, p. 330.

[187] Ibid. pp. 307-331. A review of Klerman's 19 approaches is beyond the scope of this report.

[188] Picciotto, Robert. *Evaluation Independence at DFID: An Independent Assessment Prepared for the Independent Advisory Committee for Development Impact.* August 29, 2008, p. 5, http://www.wipo.int/about-wipo/en/oversight/evaluation/pdf/iadci_dfid.pdf.

[189] Ibid.

[190] Ibid, p. 310.

[191] His 19 suggestions are grouped into three categories: (1) approaches when a funder retains the rights to specify the final text, specify the report release, and control discussion of the process; (2) approaches that limit the right to specify the final text; and (3) approaches that limit the right to specify report release. A review of Klerman's suggestions is beyond the scope of this report.

[192] DOE EERE, *Program Evaluation: Lessons Learned* (Lessons Learned from Outcome/Impact Evaluation Studies, Item #4), http://www1.eere.energy.gov/ba/pba/program_evaluation/m/lessons_learned.html.

[193] Ibid.

[194] Reingold, JPAM, p. 938-940.

[195] Ibid. p. 940.

In: The Weatherization Assistance Program ISBN 978-1-62081-745-2
Editors: S. L. King and D. R. Griffiths

Chapter 2

RECOVERY ACT: PROGRESS AND CHALLENGES IN SPENDING WEATHERIZATION FUNDS*

United States Government Accountability Office

WHY GAO DID THIS STUDY

The American Recovery and Reinvestment Act of 2009 (Recovery Act) provided $5 billion to the Department of Energy's (DOE) Weatherization Assistance Program to help low-income families by making long-term energy efficiency improvements to their homes. The Recovery Act requires GAO to conduct bimonthly reviews of how recipients such as state-level agencies use the act's funds. As part of this review, GAO examined if the act is achieving its stated purposes. The act also requires GAO to comment and report quarterly on estimates of jobs funded and counted as full-time equivalents (FTE), as reported by recipients of Recovery Act funds. GAO examined (1) the status and use of weatherization grant program funds under the Recovery Act; (2) the challenges, if any, that recipients faced in implementing the weatherization program under the Recovery Act; (3) the extent to which the weatherization program under the Recovery Act has achieved its energy and cost savings goals; and (4) the changes, if any, over time in the quality of FTE

* This is an edited, reformatted and augmented version of the Highlights of GAO-12-195, a report to the Congress, dated December 2011.

data reported by Recovery Act recipients, particularly by program recipients. GAO surveyed the 58 state-level grant recipients of the act's weatherization funds, reviewed DOE and recipient-reported data, and interviewed state and local agency officials.

GAO makes no new recommendations in this report but provides the status of prior recommendations that remain open and not implemented from GAO's Recovery Act–mandated reports. DOE generally concurred with GAO's findings and provided clarifications, which were incorporated as appropriate.

What GAO Found

As of September 2011, the 58 state-level grant recipients were awarded approximately $4.75 billion from DOE to implement the Weatherization Assistance Program under the Recovery Act and reported spending about $3.46 billion (about 73 percent). DOE expects to meet or exceed its production target of 607,000 homes and spend most of the act's funds because some recipients have been able to exceed their production targets because of a lower average cost of weatherizing homes and lower training and technical assistance expenses than anticipated. In response to GAO's prior recommendation that DOE clarify production targets and funding deadlines, among other things, DOE officials provided documentation showing actions taken concerning targets but failed to provide clarification of the consequences for not meeting the targets.

Most recipients reported experiencing more implementation challenges in the first year of the Recovery Act than in the third year. Initial challenges included implementing new wage and reporting requirements and balancing training and technical assistance requirements with production targets. In the absence of a spending deadline for the weatherization grant program, DOE established a deadline of March 31, 2012, for recipients to complete spending Recovery Act weatherization funds. Recipients reported concerns with completing final Recovery Act requirements by DOE's deadline, and continuing to support weatherization efforts after the deadline. Officials from state and local agencies reported seeking alternative sources of funding to mitigate the loss of federal funds. DOE weatherization officials said they requested a 2-year extension from the Secretary of Energy to allow some recipients, on a case-by-case basis, to spend any remaining Recovery Act funds after March 2012. However as of November 2011, it had not been determined if an extension would be available for recipients. In the interim, the Office of Management and Budget released a September 2011 memorandum stating that Recovery Act funds should be spent by September 2013.

A long-term Weatherization Assistance Program goal is to increase energy efficiency through cost-effective weatherization work. March 2010 estimates from an Oak Ridge National Laboratory study project that energy savings will likely exceed the program's costs, so that every $1 spent on the weatherization program for 2009 through 2011 would result in almost $2 in energy savings over the useful life of the investment; the laboratory plans to issue more definitive estimates in 2013. In response to GAO's prior recommendation that DOE revisit methodologies used to determine the most cost-effective work, DOE officials stated that the results of this 2013 study will be used to strengthen current protocols for determining the most cost-effective weatherization work.

According to GAO's analysis, the quality of FTE data reported by recipients to FederalReporting.gov has improved over time. DOE performs quality assurance steps on the data that recipients provide to FederalReporting.gov, and DOE officials reported that data quality continues to improve. According to Recovery.gov, the Recovery Act funded approximately 14,090 FTEs for the quarter ending September 30, 2011. FTEs are declining since the quarter ending December 2010 as weatherization work is completed.

ABBREVIATIONS

DOE	Department of Energy
DOT	Department of Transportation
EECBG	Energy Efficiency and Conservation Block Grant
FHWA	Federal Highway Administration
FTE	full-time equivalent
HHS	Department of Health and Human Services
HUD	Department of Housing and Urban Development
LIHEAP	Low Income Home Energy Assistance Program
NEAT	National Energy Audit Tool
OHS	Office of Head Start
OMB	Office of Management and Budget
ORNL	Oak Ridge National Laboratory
PAGE	Performance and Accountability for Grants in Energy

SERC Sustainable Energy Resources for Consumers
TCAP Tax Credit Assistance Program

December 16, 2011

Report to the Congress

In response to the recent economic crisis, Congress enacted the American Recovery and Reinvestment Act of 2009 (Recovery Act) to, among other things, preserve and create jobs and promote economic recovery.[1] Since the Recovery Act was enacted, the Department of the Treasury has paid out approximately $420 billion in Recovery Act funds for use by, among others, state and local governments.[2] The Recovery Act directed states to use the funds for various purposes, including assisting those most affected by the recession and investing in transportation, environmental protection, and other infrastructure to provide long-term economic benefits. Furthermore, the Recovery Act gave preference to activities that could be started and completed expeditiously. The Recovery Act provided $5 billion to the Department of Energy's (DOE) Weatherization Assistance Program (weatherization program), which assists low-income families in reducing their energy bills by making long-term energy efficiency improvements to their homes. These improvements include installing insulation, sealing leaks, and modernizing heating equipment and air conditioning equipment. DOE distributed the Recovery Act funds for weatherization assistance to 58 recipients (state-level agencies), including all 50 states, the District of Columbia, 5 territories, and 2 Indian tribes, to be spent by March 31, 2012.

The Recovery Act requires that GAO conduct bimonthly reviews of how the act's funds are used by recipients. As part of this review, we examined whether the act is achieving its stated purposes. The Recovery Act also requires GAO to comment and report quarterly on estimates of jobs funded and counted as full-time equivalents (FTE), as reported by recipients of Recovery Act funds.[3] In this report, we update our May 2010 report and include new information on the use of Recovery Act funds provided for the weatherization program.[4] Specifically, our objectives were to examine (1) the status and use of weatherization grant program funds under the Recovery Act; (2) the challenges, if any, that recipients faced in implementing the weatherization program under the Recovery Act; (3) the extent to which the weatherization program under the Recovery Act has achieved its energy and cost savings goals; and (4) the changes, if any, over time in the quality of the

FTE data reported by Recovery Act recipients (state-level agencies), particularly by weatherization program recipients.

To address all four objectives, we conducted a web-based survey of all 58 recipients of weatherization funding under the Recovery Act and received 55 responses.[5] We also interviewed officials from DOE and national associations that focus on weatherizing low-income housing, and 10 selected state and territorial offices responsible for overseeing the weatherization program.[6] We selected these states using several criteria, including the percentage of Recovery Act funds spent, the percentage of households weatherized, the size of the Recovery Act grant, the percentage of funding received out of the total funding amount of $5 billion, and geographic location. We also conducted site visits and spoke with state officials and select local agencies in 7 states and conducted telephone interviews with officials in the other 2 state-level agencies.[7] In September 2011, we also attended the National Association for State Community Services Programs' Annual Conference in Seattle, Washington, where we met with a number of recipients and DOE weatherization officials. In addition, to determine the status and use of weatherization grant program funds under the Recovery Act, we reviewed the relevant laws, regulations, and program guidance; analyzed funding and production data provided by DOE; and spoke with officials from national associations, DOE, state weatherization offices, and select local agencies. To determine the challenges recipients faced in implementing the weatherization program under the Recovery Act, we analyzed 15 DOE Inspector General reports and 27 state auditors' reports that had examined the Weatherization Assistance Program under the Recovery Act between October 2009 and August 2011. We interviewed DOE, state, and local agency officials on challenges, if any, they faced and continue to face in implementing the program, including monitoring, training, and closing out their Recovery Act weatherization program grants. To determine the extent to which the Recovery Act's weatherization program is expected to achieve energy and cost savings goals, we reviewed Oak Ridge National Laboratory's (ORNL) 2010 cost-benefit estimates of expected costs and benefits of the weatherization program under the Recovery Act and its plan for a more detailed evaluation of impacts after the program ends in 2012. ORNL is conducting these analyses under a Recovery Act–funded DOE weatherization grant. We also interviewed the laboratory officials who prepared these documents. To determine how the quality of the FTE data has changed over time, as reported by Recovery Act recipients, we reviewed nine rounds of data reported by recipients and made available on a website created for the Recovery Act—Recovery.gov— and spoke with DOE and state weatherization officials about their processes for collecting and verifying the data they report.[8] See appendix I for a detailed discussion of our overall scope and methodology and appendix III for further information on our survey.

Our oversight of programs funded by the Recovery Act has resulted in more than 100 related products with numerous recommendations since we began reporting on the Recovery Act.[9] This report updates agency actions in response to recommendations from previous bimonthly and recipient reporting reviews that have not been fully implemented (referred to in this report as open recommendations) in appendix IV.

We conducted this performance audit from December 2010 through December 2011 in accordance with generally accepted government auditing standards. Those standards require that we plan and perform the audit to obtain sufficient, appropriate evidence to provide a reasonable basis for our findings and conclusions based on our audit objectives. We believe that the evidence obtained provides a reasonable basis for our findings and conclusions based on our audit objectives.

BACKGROUND

DOE's Weatherization Assistance Program was created under Title IV of the Energy Conservation and Production Act of 1976. By making long-term energy efficiency improvements, such as installing insulation, sealing leaks, and modernizing heating and air conditioning equipment, the weatherization program aims to, among other things, increase the energy efficiency of homes owned or occupied by low-income persons, reduce their total residential expenditures, improve their health and safety, and reduce the burden of energy prices.[10] According to DOE, such improvements allow these households to spend the money saved on energy costs for other pressing needs. DOE makes weatherization program funds available through formula-based grants to state-level agencies in the 50 states, the District of Columbia, U.S. territories, and Indian tribes.[11] State-level agencies (recipients) then contract with local agencies (also referred to as subrecipients) to deliver weatherization services to eligible residents. In addition to awarding the state-level grants, under the Recovery Act, DOE awarded Sustainable Energy Resources for Consumers (SERC) grants to 27 recipients on a competitive basis. The Energy Independence and Security Act of 2007 included a provision that in any year the weatherization program funding is at least $275 million; up to 2 percent of the funding can be apportioned for SERC grants to pay for "materials, benefits, and renewable and domestic energy technologies" that are not traditionally allowed under the weatherization program. DOE awarded SERC grants for the first time in August 2010 under the Recovery Act.

The Recovery Act appropriated $5 billion for the Weatherization Assistance Program, which represents a significant increase for a program that had received about $225 million per year in recent years. Using Recovery Act funds, DOE awarded grants that will allow local agencies to weatherize approximately 607,000 homes by March 31, 2012, according to agency officials.[12] In the absence of a spending deadline for the weatherization grant program, DOE established a deadline of March 31, 2012, for recipients to complete spending Recovery Act weatherization funds.[13] In addition to receiving Recovery Act funds, DOE continued to receive no-year weatherization funds through regular appropriations for fiscal years 2009, 2010, and 2011: DOE received $450 million in fiscal year 2009;[14] $210 million in fiscal year 2010; and $174 million in fiscal year 2011. Recipients of these funds may carry over balances from previous fiscal years. DOE guidance instructs recipients to spend their Recovery Act weatherization funds first, but DOE also encourages recipients to use their appropriations in the year received to avoid carrying over balances.

In addition to being eligible for DOE funds, states and territories are eligible to receive Low Income Home Energy Assistance Program (LIHEAP) funds administered and distributed by the Department of Health and Human Services; recipients may spend up to 15 percent of these funds on weatherization activities. The maximum amount of LIHEAP funds available to states and territories for weatherization activities was approximately $750 million in fiscal year 2009, $750 million in fiscal year 2010, and $705 million in fiscal year 2011. These funds represent a significant increase from previous years.[15]

The Recovery Act changed a few existing provisions for weatherization. Specifically:

- The income eligibility level to qualify for services increased from 150 percent to 200 percent of the poverty level.
- The statewide average allowable assistance level per home increased from $2,500 to $6,500.
- The allowable funding for training and technical assistance increased from 10 percent up to 20 percent.[16]

In May 2010, we reported that recipients varied somewhat in how they determined income eligibility. We recommended that DOE establish best practices in how income eligibility should be determined and documented.[17] In response to our recommendation, in September 2010, DOE issued guidance that clarified the definition of income eligibility and strengthened the income eligibility requirements.

The Recovery Act also applied Buy American and Davis-Bacon provisions to DOE's weatherization program for the first time. Under the Recovery Act's Buy American provision, all steel, iron, and manufactured goods used in Recovery Act-funded projects must be produced in the United States, subject to limited exceptions. Under the Davis-Bacon provision of the Recovery Act, recipients are required to pay the prevailing wage rate in the locality in which Recovery Act work is conducted, as determined by the Department of Labor.[18] Davis-Bacon provisions had not previously applied to weatherization projects funded through regular DOE appropriations.

DOE provided the Recovery Act funds incrementally, as recipients completed certain requirements. Initially, each recipient received the first 10 percent of its allocated funds, which could be used for start-up activities, such as hiring and training staff, purchasing needed equipment, and performing energy audits of homes, among other things. Before a recipient could receive the next 40 percent of its funds, the recipient had to submit a weatherization plan outlining how Recovery Act weatherization funds would be used. In this plan the recipient identified the number of homes to be weatherized and the strategies it would use to monitor its funds. To receive access to the final 50 percent of funds, a recipient had to weatherize at least 30 percent of the homes identified in its weatherization plan and meet other requirements. These other requirements included implementing the monitoring and inspection protocols established in its weatherization plan; monitoring its local agencies at least once each year to determine compliance with administrative, fiscal, and state policies and guidelines; ensuring that local quality controls are in place; inspecting at least 5 percent of completed homes during the course of the respective year; submitting timely and accurate progress reports; and conducting monitoring reviews to confirm acceptable performance.

The Recovery Act sought to increase transparency and accountability in a program's use of funds. Accordingly, DOE requires recipients to report detailed expenditure and performance information to the Weatherization Assistance Program through a web-based application—the Performance and

Accountability for Grants in Energy (PAGE) system. Recipients are to provide quarterly reports in PAGE within 30 calendar days of the end of each quarter year. The information to be reported includes hours worked, expenditures, and homes completed by each subrecipient. In addition, section 1512 of the Recovery Act requires recipients to report certain information quarterly. Recipients report, among other types of information, expenditures, the progress of their projects, and FTEs. Recipients provide the reports through FederalReporting.gov, and the data are then made available to the general public through the Recovery.gov website.

PROGRESS IN SPENDING AND PRODUCTION TARGETS

According to DOE data, as of September 30, 2011, recipients had spent most funds, primarily to weatherize low-income housing and provide training and technical assistance to train state and local agency officials as well as private contractors. In addition, DOE weatherization officials told us that the department is on track to meet or exceed its production target of weatherizing 607,000 low-income homes by March 2012,[19] but some individual recipients may not meet the production targets outlined in their state plans.

Recipients Spent Most Funds Primarily to Weatherize Low-Income Housing and Provide Training and Technical Assistance

As of September 30, 2011, recipients had been awarded approximately $4.75 billion for implementing the weatherization program under the Recovery Act and spent about $3.46 billion, or nearly 73 percent, of this total, according to DOE data. Figure 1 shows how DOE allocated the funds provided by the weatherization program under the Recovery Act, and DOE and recipients' expenditures as of September 2011.

Recipients spent about 88 percent of their funds—about $3.04 billion—on weatherizing about 563,000 low-income homes. These expenditures were for labor and materials, transportation of workers and materials to the job sites, and the purchasing of vehicles.[20] Figure 2 shows two examples of weatherization work performed on homes in Maine and Washington states.

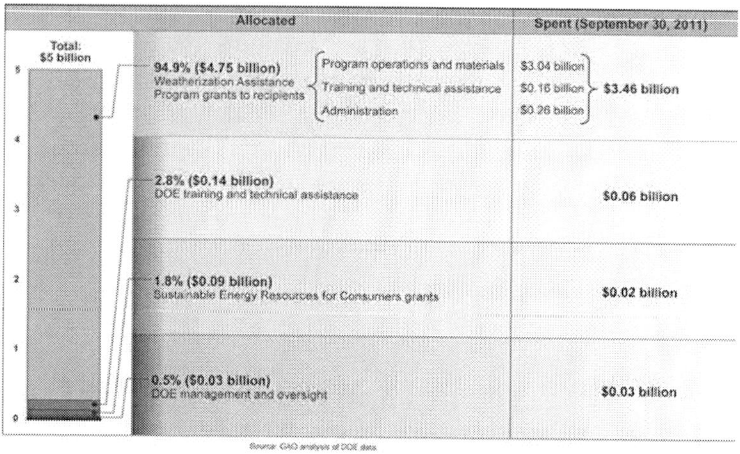

Figure 1. DOE Allocation of the $5 Billion Recovery Act Weatherization Assistance Program Appropriation and DOE and Recipient Expenditures, as of September 30, 2011.

Figure 2. Insulation Installed in the Basement of a House in Maine and beneath a House in Washington State.

As figure 1 also shows, for training and technical assistance, recipients spent about $0.16 billion of their total allocation, as of September 30, 2011. According to DOE guidance, training and technical assistance expenditures are intended to maintain the efficiency, quality, and effectiveness of the weatherization program. The recipients used these funds to train state and local agency officials, and private contractors. For example, South Carolina reported that the state used its training and technical assistance funds to work with six technical colleges in the state to set up classes, such as Hot Climate Whole House Weatherization—Site-Built and Hot Climate Whole House Weatherization—Mobile-Home. From April 2010 through March 2011, over 1,500 participants received training in these classes. Figure 3 shows a South

Carolina weatherization test house used in training. Furthermore, according to Oregon officials we interviewed, the Recovery Act technical assistance and training funds allowed them to administer the weatherization program at two tribal agencies. Before the Recovery Act, these officials told us, the state did not have sufficient funds to initiate a weatherization program with those tribes. With the Recovery Act funds, Oregon officials said they were able to spend a month with each tribal agency to help set up the weatherization programs.

For program administration, recipients spent about $0.26 billion, according to DOE data. This amount is about 7.5 percent of total recipient expenditures, or less than the 10 percent that DOE weatherization program regulations generally allow for administrative purposes. Administrative costs include salaries and fringe benefits for fiscal and administrative staff.

As figure 1 also shows, DOE allocated about $0.25 billion of the $5 billion in weatherization funds under the Recovery Act for other weatherization purposes, including DOE training and technical assistance, SERC grants, and DOE management and oversight. DOE's weatherization training and technical assistance plan for the Recovery Act noted that the expansion of the weatherization program under the Recovery Act necessitated a significant growth in training capacity. To that end, the plan outlined a range of program operation activities: training, including developing tools and information resources; and technical assistance, including measuring and documenting performance, monitoring programs, and promoting advanced techniques and collaborations to further improve program effectiveness. Table 1 outlines some of the training and technical assistance efforts that DOE undertook.

Source: GAO.

Figure 3. Weatherization Test House at a Training Center in South Carolina.

Table 1. Funding Allocation for Selected DOE National Training and Technical Assistance Initiatives

Recipient(s)	Training and technical assistance activity	Allocation
34 weatherization training centers in 27 states	Develop and expand weatherization training centers	$28,760,810
Energy Enterprise Solutions	Develop the National Platform for Weatherization Training to provide interactive online weatherization training	19,658,348
Oak Ridge National Laboratory	Conduct the national evaluation of the weatherization program under the Recovery Act, including estimate of energy savings and cost effectiveness	18,836,000
Oak Ridge National Laboratory and the Institute of Building Technology and Safety	Conduct quality assurance program to ensure proper use of weatherization grants by recipients	10,000,000
National Association for State Community Service Providers	Support DOE's Weatherization Plus Health initiative, including partnership development and regional weatherization training[a]	8,495,293
National Institute of Building Sciences	Develop multimedia modules for training weatherization technical auditors and installers	7,000,000
Data Tree, Inc., of Virginia	Develop the PAGE module for the weatherization program	$4,795,327

Source: GAO analysis of DOE data.

[a] DOE's Weatherization Plus Health initiative is a national effort to coordinate resources to improve the energy efficiency, health, and safety of low-income homes.

As we reported in May 2010, worker training and certification requirements varied among the recipients.[21] We recommended that DOE accelerate its efforts to develop national standards for weatherization training, certification, and accreditation. DOE stated that it, along with the Department of Labor, released for comment a draft—"Workforce Guidelines for Home Energy Upgrades"—for single-family homes in November 2010. DOE weatherization officials noted that they received nearly 1,000 comments to the draft guidelines, which will likely delay publication of the final product.

Source: GAO.

Figure 4. Solar Panels Installed at a SERC Project in Massachusetts.

In August 2010, DOE awarded SERC grants to 101 local agencies in 27 states to install, test, and report on renewable and domestic technologies not currently covered by the weatherization program. Such technologies include solar photovoltaic systems, tankless or on-demand hot water systems, and geothermal heat pumps. As of September 30, 2011, the 27 states had spent about $16 million of their SERC grants. Figure 4 shows the installation of solar panels on a SERC project in Massachusetts.

The Recovery Act sets aside up to 0.5 percent of the amount appropriated for the Weatherization Assistance Program for program management and oversight. As such, DOE allocated 0.5 percent of the $5 billion weatherization appropriation, or $25 million ($0.03 billion), for this purpose. According to DOE weatherization officials, these funds were transferred to DOE's Recovery Act administration account with all similar contributions from other programs to support DOE-wide Recovery Act activities. As a result, the weatherization officials could not provide specific information on how these funds were spent. In general, they said the department used these funds to add staff in DOE to enhance the program's capacity to manage the $5 billion Recovery Act appropriation.

DOE Expects to Meet Its Overall Production Target

According to DOE weatherization officials, the Weatherization Assistance Program under the Recovery Act will meet or exceed its production target of weatherizing 607,000 homes by DOE's March 2012 grant deadline because some recipients exceeded their target even as others have fallen short. As shown in figures 5 and 6, some recipients greatly increased their rate of production and expenditures over the course of the Recovery Act grant period, but progress has been uneven. Figures 5 and 6 show the percentage of planned weatherization production completed and percentage of total allocation expended, respectively, by each recipient by September 2009, September 2010, and September 2011.

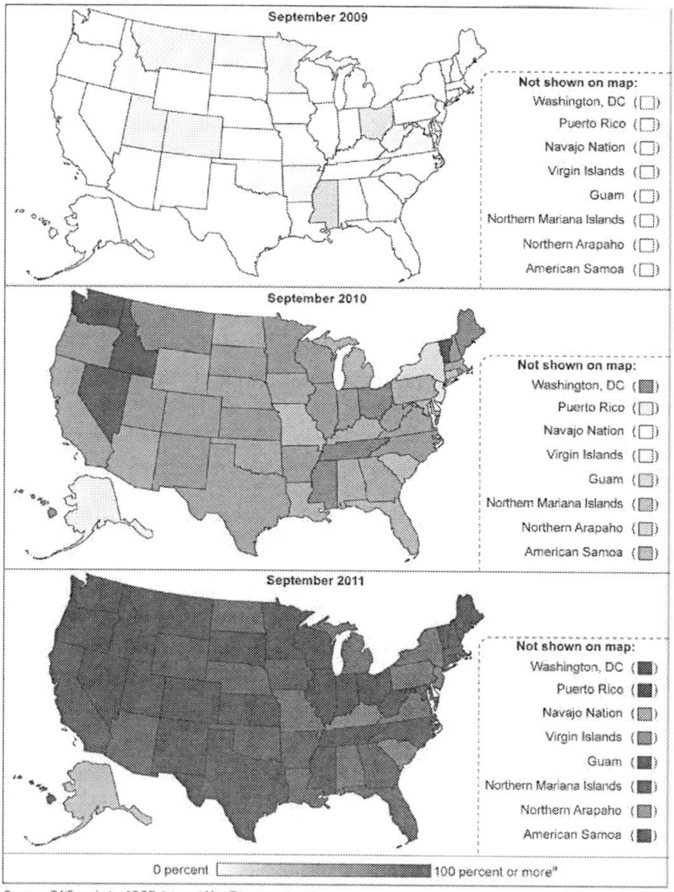

Sources: GAO analysis of DOE data and Map Resources (map).

Note: Some recipients have exceeded their production targets.

Figure 5. Percentage of Planned Weatherization Production Completed by Each Recipient, September 2009, September 2010, and September 2011.

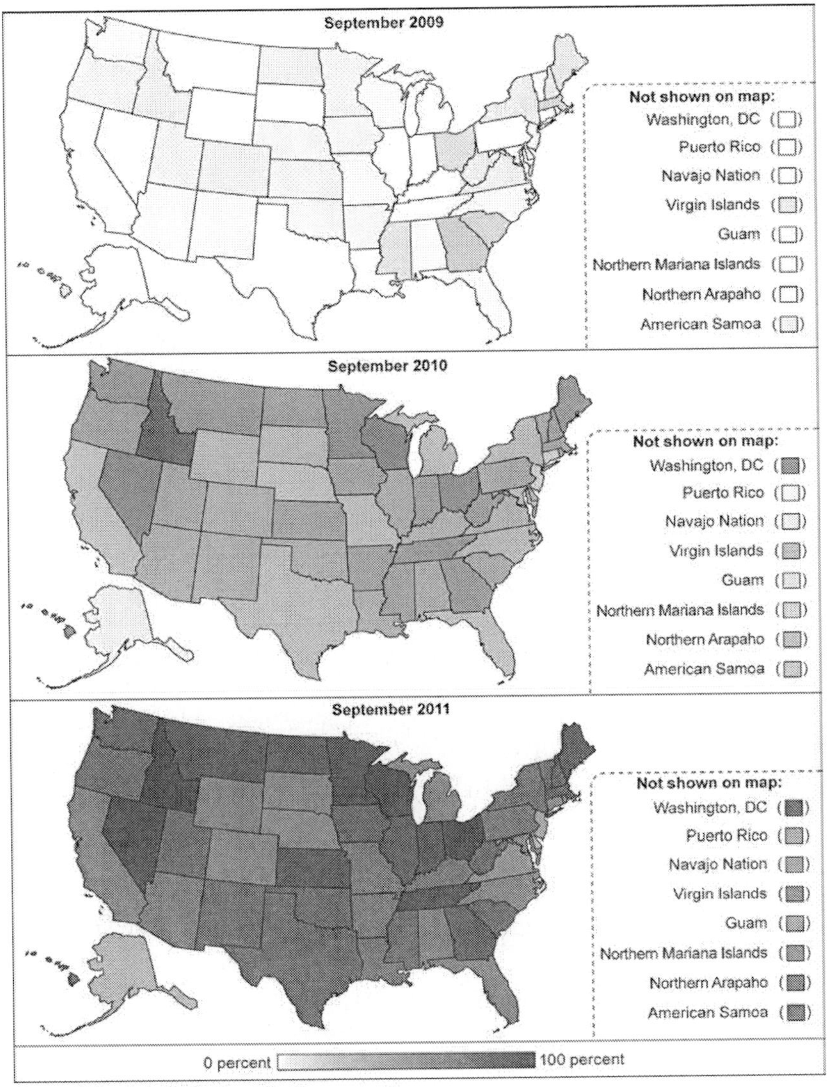

Sources: GAO analysis of DOE data and Map Resources (map).

Figure 6. Percentage of Total Allocation Expended by Each Recipient, September 2009, September 2010, and September 2011.

DOE will likely meet or exceed its production target because some recipients have been able to exceed their production targets for the following reasons:

- *The average cost of weatherizing homes was lower than anticipated.* Since 2001, the statewide average expenditure limit per home for DOE weatherization was about $2,500, but adjusted annually to reflect changes in consumer prices. The Recovery Act increased this limit to $6,500, and in their state plans, recipients use the average cost per home of $6,500 to calculate their planned production. As of September 2011, the actual average cost per home was about $4,900, according to DOE data. As a result, some recipients may be able to weatherize more homes than originally planned. Nebraska officials, for example, cited the current economy as a driving factor for the lower cost of weatherization services. As we reported in May 2010, states were using a variety of methods to calculate the $6,500 maximum average limit per house.[22] In that report, we recommended that DOE develop and clarify weatherization program guidance that clarifies the specific methodology for calculating the average cost per home. In September 2011, DOE issued guidance that clarifies the methodology for calculating the average cost per unit weatherized and further defines the allowable cost categories used in calculating the average cost per unit. Additionally, Washington state officials attributed lower weatherization costs to their ability to weatherize more homes for less money when weatherizing multifamily units. For example, they said that when completing weatherization work in multifamily housing, such as caulking wall joints in the attic, the costs are spread out across more units, thereby reducing the cost per unit and hence the overall weatherization cost. We reported in May 2010 that some recipients had limited expertise with weatherizing large multifamily buildings. Some state officials told us that until the Recovery Act, they did not have sufficient funds to justify weatherizing multifamily buildings. Despite this unfamiliarity with weatherizing multifamily units, we reported that some recipients may feel compelled to focus on multifamily houses as a way to quickly increase their production numbers. We recommended in our May 2010 report that DOE consider and address how the weatherization program guidance is affected by the introduction of multifamily units.[23] In response to our recommendation, DOE issued additional guidance in December 2010 to the recipients regarding multifamily housing, including describing eligibility requirements for multifamily housing.

- *Training and technical assistance expenses were less than originally anticipated.* DOE's initial allocation of $0.83 billion (nearly 17 percent) for training and technical assistance to the recipients represented a substantial increase over past funding under the weatherization program funded with regular appropriations. By the second year of the Recovery Act, DOE noted that many recipients believed that they had sufficiently built up their programs, and DOE provided guidance to allow the recipients to transfer training and technical funds into program operations. As of September 2011, the recipients had transferred about $0.33 billion from their training and technical assistance allocations, increasing their program allocations to $4.25 billion. As a result, some recipients now had more funds to weatherize homes. As DOE's March 2012 deadline for spending Recovery Act weatherization funds approaches, DOE weatherization officials anticipate that more recipients will transfer training and technical assistance funds to program funds.

Some recipients may not achieve their spending and production targets. As previously discussed, when DOE obligated $4.75 billion in formula funds to the states, territories, and tribes, it released the funds to the recipients incrementally as they met certain requirements. By the end of December 2009, DOE had approved the weatherization plans for all 58 recipients, and all recipients had received access to at least 50 percent of their funds. Furthermore, as of September 30, 2010, 26 of the 58 recipients had not received access to their remaining funds; and as of September 30, 2011, with only 6 months to the March 2012 grant deadline, 6 recipients had not been granted full access to the remaining 50 percent of funds.[24]

In our May 2010 report, we found that there was some confusion among some states and local agencies regarding production targets, funding obligations, and the associated consequences of not meeting production and funding targets.[25] We recommended that DOE clarify its production targets, funding deadlines, and associated consequences while providing a balanced emphasis on the importance of meeting program requirements. In response, DOE weatherization officials produced documentation showing extensive communication with the recipients over production and expenditures. However, they did not provide documentation showing how they clarified the consequences for not meeting these targets.

Recipients Reported Fewer Implementation Challenges over Time

A majority of recipients responding to our survey reported initial challenges in implementing the weatherization program under the Recovery Act, but the percentage continuing to report these challenges had declined by the third year of the program. However, most recipients reported concerns about completing Recovery Act weatherization requirements and continuing to support weatherization efforts.

Implementation Challenges Declined over Time

In the first year of the weatherization program funded under the Recovery Act, most recipients who responded to our survey reported facing challenges in 26 of the 29 program implementation areas identified. Table 2 shows these 29 implementation areas and recipients' responses to our survey in the first and third program years.

Table 2. Percentage of Recipients Responding to Our Survey That Identified Challenges in the First and Third Years of the Weatherization Assistance Program under the Recovery Act

Program implementation area	Percentage of recipients responding that identified an area as a strong challenge or somewhat of a challenge	
	First program year	Third program year
Balancing the requirements to meet production targets and to ensure the work is done correctly	91	68
Meeting production targets	91	57
Having additional reporting requirements	90	75
Implementing new program requirements that were developed as the program progressed	89	63
Implementing Davis-Bacon requirements	88	39
Balancing training and technical assistance requirements with production targets	88	63
Adjusting to changes in existing reporting requirements	87	44

Program implementation area	Percentage of recipients responding that identified an area as a strong challenge or somewhat of a challenge	
	First program year	Third program year
Meeting federal reporting requirements (FederalReporting.gov and PAGE)	84	63
Meeting certification and training requirements	83	57
Lacking expertise weatherizing multifamily units	82	64
Having too few monitoring staff	81	40
Having insufficient number of administrative staff	76	40
Having additional reporting requirements for monitoring	75	47
Lacking expertise in weatherizing multifamily units	74	60
Conducting the required number of monitoring visits	73	42
Adjusting to changes in existing reporting requirements for monitoring	72	48
Ensuring consistency in monitoring results	71	50
Lacking expertise in monitoring multifamily units	71	47
Complying with historic preservation requirements	71	35
Lacking trained monitoring staff	69	26
State-level auditors lacking training or expertise	63	22
Lacking staff to appropriately handle external monitoring (DOE-Inspector General, state-level auditors, consultants)	63	57
Determining how to measure long-term energy savings	60	51
Meeting state's requirements to monitor subgrantees	58	29
Lacking infrastructure at state level to handle the large influx of money from the Recovery Act	57	36
Determining how to measure cost savings (i.e., dollar savings)	55	38
Planning for "cliff effect" (i.e., decline in funding after Recovery Act funding expires in March 2012)	34	89
Determining and documenting income eligibility	25	17
Determining methodology for calculating $6,500 maximum average cost per home	18	4

Source: GAO analysis of survey responses.

Note: Recipients that responded to the survey could identify each implementation area as either a strong challenge, somewhat of a challenge, or not a challenge.

As the table shows, all but 3 areas were identified as a challenge in the first year of the program by more than 50 percent of the recipients who responded to our survey.[26] Over time, however, the percentage of recipients identifying these areas as a challenge declined, although the amount of the decline varied considerably. In addition, DOE weatherization officials and recipients stated that these challenges initially delayed the weatherization of homes and spending of Recovery Act funds. By early 2011, however, having addressed some of these challenges, some recipients reported that they were weatherizing more homes in 1 month than they had previously weatherized in 1 year. As previously mentioned, however, following the initial delays common to many recipients, recipients' progress in weatherizing homes and spending Recovery Act funds has been uneven. For example, as of September 2011, Idaho had weatherized about 128 percent of its planned houses and spent nearly 99 percent of its total Recovery Act allocation, while Alaska had weatherized about 28 percent and spent about 31 percent.

Among the challenges that declined the most from the first year to the third year was implementing Davis-Bacon requirements. As table 2 shows, 88 percent of the recipients responding to our survey identified implementing Davis-Bacon requirements as a challenge in the first year of the program, but only 39 percent continued to do so by the third year.[27] According to DOE, state, and local officials, after the Department of Labor made the final wage determination for weatherization workers and as the program progressed, the process of complying with Davis-Bacon provisions became routine, and the challenge decreased for many recipients. As we have previously reported, the Recovery Act applied Davis-Bacon provisions to the weatherization program for the first time in 2009.[28] State and local officials delayed weatherization contracts because they were concerned that wage rates for weatherization had not yet been determined by the Department of Labor, and they therefore waited for the department to determine county-by-county prevailing wage rates.[29] In addition, some recipients reported facing challenges in complying with the Davis-Bacon requirement for local agencies to submit weekly certified payroll records for workers paid under Davis-Bacon wage provisions. Some recipients told us that the requirement to certify payroll discouraged some local agencies from participating. For example, in Maryland, state officials told us that this requirement was a considerable challenge because many local agencies handling weatherization work are small organizations with limited staff and therefore may face difficulty in meeting the requirement to certify payroll every week.

However, large numbers of survey respondents reported that their initial year challenges continued into the third year of the program. For example:

Having additional federal reporting requirements. As table 2 shows, 90 percent of recipients responding to our survey identified having additional federal reporting requirements as a challenge in the first program year, and by the third year 75 percent still found this area to be a challenge.[30] For example, under the Recovery Act, recipients are to report program information quarterly, such as the amount of funds spent, and full-time jobs created, to FederalReporting.gov and to DOE. Oregon officials told us meeting these requirements was challenging initially because state and local agencies had to learn how to manage, collect, and submit data. Additionally, state auditor and DOE Inspector General reports identified challenges for some recipients meeting federal reporting requirements. For example, the North Carolina State Auditor reported in March 2011 that the state did not have effective internal controls—such as internal data review procedures to ensure the accuracy of submitted data—to ensure compliance with federal reporting requirements, specifically with submitting quarterly reports to FederalReporting.gov.[31] According to DOE weatherization officials, as the program continued and reporting became routine, recipients began to meet federal reporting requirements and faced minimal challenges.

Adjusting to changes in existing reporting requirements for monitoring. Seventy-two percent of the recipients responding to our survey initially identified this implementation area as a challenge, as table 2 shows; by the third year, 48 percent of recipients responding to our surveyed still identified this as a challenge.[32] Recovery Act guidance directed federal agencies to take steps beyond standard practice to ensure compliance with grant rules and regulations. As a result, DOE enhanced its monitoring requirements for the Weatherization Assistance Program under the Recovery Act. For example, DOE's weatherization project officers track recipients' spending and weatherization of homes through the quarterly information recipients submitted to the PAGE system. The DOE Inspector General and state auditors identified problems with how recipients reported monitoring activities. For example, a DOE Inspector General report found that Virginia did not accurately report information on the number of weatherized homes reported by the local agencies to the state. The state's quarterly monitoring report to DOE indicated that 316 homes had been weatherized from July 2009 through September 2009, but the local agency, which keeps its own records, showed that 978 homes had been weatherized.[33]

The need to develop new systems for monitoring and reporting has also yielded benefits, according to DOE program officials and some recipients we interviewed. DOE weatherization officials told us that monitoring and reporting have made the weatherization program stronger by identifying deficiencies and other weaknesses and having the department resolve them. Recipients also reported that the challenge of meeting federal reporting requirements provided an opportunity to improve their programs. These improvements included replacing or enhancing their data collection systems, creating specific audit tools for weatherization work, and developing manuals on best practices. For example:

- Officials in Puerto Rico told us that they developed their own specialized audit tool based on the original tool developed by Oak Ridge National Laboratory for the weatherization program. Puerto Rico's specialized tool focuses on the relevant weatherization activities—such as replacing refrigerators, water heaters, and light bulbs—to address the territory's year-round tropical conditions in which, officials told us, the coldest temperature of the year typically reaches 60 degrees Fahrenheit.
- In Maine, officials identified best practices in the weatherization program and created a manual on weatherization procedures that provides additional guidance on the state's program.

Balancing training and technical assistance requirements with production targets. As table 2 shows, 88 percent of recipients responding to our survey initially reported this implementation area as a challenge, and by the third year of the program, 63 percent of recipients still reported that this was a challenge.[34] According to DOE, specialized knowledge and skills are required at the federal, state, and local levels to run an effective weatherization program; these include core competencies in areas such as evaluating buildings to determine weatherization needs, installing weatherization measures, and conducting final inspections of weatherization projects. For recipients to have enough weatherization workers with the specialized knowledge and skills, some recipients focused on increasing their training capacity with Recovery Act funding. According to these officials, many local agencies initially rushed to meet their monthly production targets, and the weatherization work suffered, requiring local agencies to conduct rework. Following the increase in training and experience, state officials found that local agencies have to redo less work. In Washington state, officials faced less of a challenge with balancing training with production targets because officials hired staff who had completed weatherization training previously and therefore could complete weatherization projects instead of focusing on meeting training requirements.

Many recipients cited the amount of training weatherization workers received as a positive and lasting impact of the Recovery Act. New weatherization workers were hired to meet the increased production targets under the Recovery Act and as a result received a substantial amount of training. For example, South Carolina officials told us that the training provided through the Recovery Act–funded weatherization program helped individuals gain new skills that are transferable to other areas, such as the private homebuilding industry and local utility energy auditing profession, and will increase their ability to obtain employment in the future. In addition, before the Recovery Act, Puerto Rico had not received weatherization program funding, and territorial officials stated that they used training and technical assistance funds to develop a training center, which will provide weatherization training to individuals in Puerto Rico and other neighboring Caribbean islands.[35] The officials also stated that the center will be sustained with regular appropriations after the Recovery Act funding has ended.

Our interviews with recipients and our reviews of reports by the DOE Inspector General and state auditors, along with our survey data, showed that there were other notable challenges faced by recipients in implementing their programs:

- *Building operational capacity.* Building the operational capacity of weatherization programs to accommodate greater funding levels—some of which were more than 10 times larger than their regular weatherization appropriations—was an initial challenge recipients faced. With funding increased to $5 billion over 3 years under the Recovery Act, recipients needed to scale up their operations—hiring and training staff and monitors, renting new facilities, and purchasing more vehicles and equipment—to handle the influx of funding and meet their individual state targets to weatherize homes and spend Recovery Act funding by the March 2012 deadline. For example, the funding Florida received under the Recovery Act for weatherization was substantially more than the average funding it had received for the weatherization program supported with regular appropriations.[36] Florida officials told us that they welcomed the opportunity to weatherize thousands of additional homes with the higher funding level under the Recovery Act, but it was difficult for the state to increase its capacity for the size and scope of the program. Similarly, funding for Georgia's weatherization program under the Recovery Act was over 10 times larger than the average funding for the state's weatherization program supported with regular appropriations. In response, Georgia contracted with the University of Georgia Cooperative Extension to help manage the increased amount of state-level monitoring necessary for the weatherization program under the Recovery Act. Georgia officials stated that the University of Georgia hired 24 monitors to conduct monitoring. State officials told us that before the Recovery Act, state-level monitors conducted on-site monitoring at local agencies about once a year, while under the Recovery Act and using the University of Georgia monitors, the local agencies are visited by a monitor at least once a month.

- *Monitoring deficiencies.* Several DOE Inspector General and state auditor reports identified state-level deficiencies in monitoring efforts. For example, the DOE Inspector General identified significant problems in Illinois with on-site monitoring, inspection, and compliance.[37] Additionally, 14 of the 27 state auditor reports we reviewed identified deficiencies with state-level monitoring of local agencies. For example, the Missouri State Auditor reported that the state had failed to review subrecipient expenditures to ensure costs were allowable, did not have policies and procedures to ensure review of expenditures in monitoring, and did not adequately document subrecipient compliance with federal and state procurement requirements.[38] In May 2010, we recommended that DOE set time frames for states to develop and implement state monitoring programs.[39] As of September 2011, DOE had taken limited action to address this recommendation, and it remains largely unaddressed.[40]
- *Internal controls.* Several DOE Inspector General and state auditor reports identified issues with internal controls. For example, the DOE Inspector General identified numerous internal control challenges for West Virginia, including the lack of adequate financial controls to ensure the accuracy and integrity of financial information and costs incurred.[41] Similarly, a New Jersey state auditor found that the state was not complying with federal regulations on how to handle vehicle and equipment purchases over $5,000 that were made with Recovery Act funds.[42] We reported previously that the extent of internal controls varied greatly across some recipients.[43] We recommended

that DOE, in conjunction with both state and local weatherization agencies, develop a best practices guide for key internal controls that should be present at the local weatherization agency level to ensure compliance with key program requirements. In response to our recommendation, DOE has taken some action, including issuing a memo in May 2011 to all weatherization recipients outlining key internal control requirements.[44] DOE weatherization officials told us they did not intend to issue a best practice guide for key internal controls because they believed there were sufficient documents in place to require internal controls, such as grant terms and conditions, and a training module available on the Weatherization Assistance Program Technical Assistance Center website.[45]

Challenges Reported for Completing Recovery Act Requirements and Continuing Weatherization Efforts

DOE program and state officials we interviewed reported some challenges as they approach DOE's deadline of March 31, 2012, for completing their use of Weatherization Assistance Program grants under the Recovery Act. These challenges included (1) satisfying DOE's requirement to finish spending Recovery Act funds if the March 2012 deadline is not extended[46] and (2) finding alternative funding sources for the weatherization program after March 2012.[47]

Satisfying DOE's requirement to finish spending funds. Some recipients will still have Recovery Act weatherization funds remaining after the March 2012 deadline, even if they have already met their production targets, according to DOE projections based on data as of September 2011. DOE program officials projected that at least 5 percent, or more than $300 million, of Recovery Act program and training and technical assistance funds may remain unspent by more than 30 recipients. The amount of funds remaining varies greatly among recipients. For example, DOE projections indicate that both Minnesota and Oregon will meet their production targets by March 2012, but the projections also indicate that Minnesota may have a $148,000 balance and Oregon a $5 million balance. If these remaining funds were to be spent, recipients might be able to weatherize an estimated additional 50,000 to 60,000 homes, according to DOE projections.

In order to expend any remaining Recovery Act funds, DOE program officials told us that they requested an extension from the Secretary of Energy to allow recipients to spend their remaining Recovery Act weatherization funds until September 2013, rather than the March 2012 deadline the department originally set.[48] As of November 2011, it had not been determined if an extension would be available for recipients. DOE weatherization officials told us if an extension is granted, it will be available on a case-by-case basis, using criteria such as the amount of funds remaining, and the rate of expenditures, to determine the recipients who will receive extensions as well as the length of those extensions.

Finding alternative funding sources. As they enter the third and final year of the weatherization program under the Recovery Act, 89 percent of the recipients responding to our survey identified the expected decline in funding after the end of Recovery Act funds as a challenge.[49] As a result, they will need to find alternative sources of funding for their weatherization efforts. DOE program and state officials also told us that without alternative sources of funding, many recipients will be unable to continue providing weatherization services at the current production level. In addition, as the Recovery Act funding has begun to be completely expended, many recipients have begun to lay off weatherization employees. For example, DOE program officials told us that Illinois had spent approximately 82 percent of its Recovery Act grant funds as of September 30, 2011, and had laid off approximately 1,000 employees. Washington state officials also told us that they had hired 15 new staff and 5 inspectors, to help achieve the state's Recovery Act weatherization targets. However, the five inspectors were no longer working for the state weatherization program by May 2011, when it met its Recovery Act production target. Georgia officials told us that neither the state nor the local agencies will have sufficient funding to sustain their current production after their Recovery Act weatherization grant ends.

Officials from several states, local agencies, and training centers stated that they are searching for alternative funding sources in anticipation of the decrease in overall weatherization assistance funding. Recipients have received funding from several sources, such as regular DOE weatherization appropriations, LIHEAP funding, and assistance from utility companies. In particular:

- According to DOE data, all 58 recipients currently receive regular DOE weatherization appropriations, which are available until expended, and they have a combined carryover balance of approximately $317 million in these appropriations that they can spend after March 31, 2012.[50] Some recipients are currently using their regular DOE weatherization appropriations. Recipients such as Idaho and Washington state, which have already spent more than 75 percent of their Recovery Act weatherization grants, are currently using their regular DOE weatherization appropriations and therefore may have fewer funds to carry over and spend after March 31, 2012. In addition, 74 percent of those we surveyed reported that they received LIHEAP funds.[51]

- Weatherization funds from nonfederal sources, such as local utility companies, are also decreasing, according to some state program officials. In Maryland, for example, officials told us that they had previously had two nonfederal sources of weatherization funds: the Regional Greenhouse Gas Initiative and the Electric Universal Service Program. However, because the Regional Greenhouse Gas Initiative has generated only limited revenue recently, no funding has been allocated for the weatherization program through this initiative.[52] The state has not received funding through the Electric Universal Service Program for the past 2 years.[53] Officials in Georgia stated that the grant they received from one utility company ended in May 2011, leaving them with funding from one other nonfederal entity. Two state officials we spoke with, from Rhode Island and Massachusetts, told us that they have sufficient funds from utility companies to compensate for the end of their Recovery Act weatherization grants.

Many recipients and local agencies we spoke with are identifying new sources of funding and strategies to continue providing weatherization services after Recovery Act funding ends. For example, Baltimore Housing officials told us that they are trying to combine weatherization measures along with, for example, health and safety measures in one project so that they can draw upon multiple revenue sources. At training centers in Georgia and Maryland, officials told us that they are looking to fund their programs by converting some of their weatherization courses into fee-based training. Also, in Maryland, the training center officials stated that they will collaborate with a local community college to offer an academic degree in weatherization.

Early Estimates Indicate Cost-Effective Energy Savings

A long-term goal of the weatherization program is to increase energy efficiency through cost-effective weatherization work, and according to the Oak Ridge National Laboratory's March 2010 preliminary estimates, the Recovery Act weatherization program is expected to result in energy savings whose value will significantly exceed program costs.[54] ORNL plans to issue a report with more definitive estimates of the benefits and costs of the program in 2013 that describes the results of its extensive evaluation of the weatherization program under the Recovery Act. For this purpose, the ORNL plan includes an extensive data collection effort on program costs and impacts on energy consumption for residential units weatherized during the program years 2009 through 2011.[55]

March 2010 Preliminary Estimates

According to ORNL's March 2010 preliminary estimates, every $1 spent on the weatherization program for 2009 through 2011 would result in $1.80 in energy savings over the useful life of the investment.[56] Because of the prospective nature of the March 2010 analysis—preceding the program's actual implementation and results—ORNL relied on estimates of the costs of weatherization investments and energy savings to be expected over the 3 years covered by the Recovery Act; these estimates were generated using ORNL's computer-based program, the National Energy Audit Tool (NEAT). The detailed ORNL estimates reflect the differences among states in energy consumption for heating and cooling and in the sources of energy used in the residential sector. For example, these differences include the following considerations:

- Energy consumption for home heating in New England is generally much higher than in the Southwestern states of Arizona and New Mexico, where energy consumption for air conditioning is relatively higher.
- Home heating in the Midwest relies heavily on natural gas, which is more economical for this purpose than electricity, which is more heavily used in the Southeast.

- Electric utility rates in New England are, on average, considerably higher than in the Mountain states.

As a result of these differences, the estimated energy savings and the dollar values associated with these savings in the ORNL analysis vary widely from region to region.

ORNL's March 2010 estimates also included estimates of program savings in electric bills from weatherization projects to reduce energy consumption by installing more efficient appliances. However, because ORN had more limited information on energy consumption by residential electric appliances, it simplified its assumption. It assumed that energy savings from improved efficiency would result in a uniform savings of 870 kilowatt-hours per year regardless of location.[57]

ORNL's March 2010 estimates also accounted for geographic differences in the cost of weatherizing homes. For example, the cost used in the model for weatherizing a home in New York City is about 15 percent higher than for a home in Charleston, South Carolina. The difference reflects the need for different weatherization projects for the two locations, as well as differences in the costs of labor and materials. Table 3 shows some of the principal factors considered and the resulting estimates of savings and costs in ORNL's 2010 preliminary estimates.

Table 3. ORNL's 2010 Preliminary Estimates of Energy Savings and Costs under the Recovery Act's Weatherization Assistance Program

Principal factors considered (per household)	Estimates
Average heating and cooling bill reduction for the year 2010	$440
Savings in electricity bills for electric appliances (excluding heating and cooling) for 2010	$100-$180
Average cost of weatherizing a housing unit under the program	$5,760
Average value of energy saving	$10,350
Energy benefit cost ratio (benefit of $10,350 divided by cost of $5,760)	1.8

Source: ORNL.

Note: All dollar values were converted from 2008 dollar values in the original ORNL report to 2010 dollar values using the "Calendar Year Chain-Weighted GDP Price Index," and we rounded the figures.

In its preliminary estimates, ORNL also examined the nonenergy benefits of the weatherization program under the Recovery Act, such as income generated from indirect employment,[58] reduced emissions of carbon dioxide and pollutants resulting from lower residential fuel and electricity consumption, and improved health and safety for program recipients. According to ORNL's estimates, these nonenergy benefits add an additional $0.70 in nonenergy benefits for every dollar spent on the program, or about $4,200 for each home weatherized.

It is beyond the scope of our review to evaluate the reliability of ORNL's 2010 preliminary estimates, but their prospective nature as projections means that there is inherent uncertainty about how well they predict actual future outcomes. As mentioned above, ORNL's 2010 analysis relied on estimates of weatherization investment costs and expected energy savings generated using ORNL's computer-based program, NEAT; and the ORNL report mentioned that the analysis adjusted the NEAT energy savings downward because of evidence that NEAT tends to overestimate energy savings. ORNL told us, however, that the use of the NEAT energy estimates is one of many uncertainties surrounding the preliminary estimates. In contrast, the forthcoming 2013 report will rely on extensive data from household units actually weatherized over the period 2009 through 2011.

ORNL's 2013 Evaluation

DOE has contracted with ORNL to conduct an in-depth evaluation of any energy cost savings achieved through the Recovery Act's weatherization program for homes weatherized across the nation and in U.S. territories from 2009 through 2011. According to ORNL's plan for carrying out this evaluation, ORNL expects to issue its analysis in 2013. ORNL's analysis will be based on an extensive data collection effort. We reviewed the energy cost-effectiveness portion of this evaluation plan and found it to be methodologically sound.[59]

This evaluation of the Recovery Act weatherization program impacts is a more extensive effort and likely to lead to more definitive estimates of costs and benefits than the preliminary, prospective March 2010 estimates, according to our review of ORNL's study plan. With a budget of $19 million to complete its evaluation, ORNL plans to use more researchers and collect extensive data. The data collection will include a large sample of homes weatherized and cover recipients across the United States and its territories.

For example, ORNL will collect data on natural gas and electric utility billing histories and use data from surveys of homes weatherized in 2007 and 2008 to estimate Recovery Act period savings for homes that use fuel oil or propane for heating. State and local government officials involved in administering the program will also be surveyed, and various federal government databases will be used.

In addition, ORNL's 2013 evaluation will be based on an analysis of homes actually weatherized by each state. Because the costs and benefits of weatherization vary by region, the number of homes completed and the distribution of these homes among the regions directly influence total costs and benefits. The March 2010 estimates avoided this issue and estimated the costs and benefits of a "weighted average" of single homes nationwide, where the weighting was based on each recipient's share of total program allocations. In contrast, the forthcoming evaluation will be based on data on the number of homes actually weatherized by the individual recipients, reflecting differences in performance among them; and it will produce regional, as well as national, estimates of program costs and benefits.

In our May 2010 report, we found variation in how some local officials considered cost-effectiveness in determining what weatherization work should be performed.[60] We recommended that DOE revisit the methodologies used in determining the weatherization work that should be performed based on consideration of cost-effectiveness and develop standard methodologies that ensure that priority is given to the most cost-effective weatherization work. In response to our recommendation, DOE weatherization officials cited ORNL's ongoing 2013 evaluation. Officials told us the study results will be used to strengthen current protocols for determining weatherization work for both single and multifamily homes and to ensure cost-effectiveness.

Quality of FTE Data Has Improved over Time, and the Number of FTEs Is Declining

To meet our Recovery Act mandate to comment on recipient reports, we continued to monitor the data recipients reported from October 1, 2009, through September 30, 2011. For this report, we focused our review on the quality of FTE data as reported by recipients of weatherization funds under the Recovery Act to FederalReporting.gov and efforts DOE made to validate those data.[61] Each quarter, DOE performs quality assurance steps on the data that recipients provide to FederalReporting.gov , and officials reported that data quality continues to improve. Through these reviews and their interactions with recipients, DOE weatherization officials concluded that recipients now understand the reporting process and stated that the data reported for each reporting period have been of higher quality than the previous period. Reviewing the weatherization program data in FederalReporting.gov from the second reporting period, which ended on December 31, 2009, to the ninth reporting period, which ended on September 30, 2011, we continued to check for errors or potential problems by repeating analyses and edit checks discussed in previous reports. We reviewed data associated with the 58 weatherization program recipient reports and 34 training center grants made to weatherization recipients posted on Recovery.gov for the ninth reporting quarter.[62]

Overall, the recipients reported on Recovery.gov that the Recovery Act weatherization program funded an increasing number of FTE positions for recipients and subrecipients from the quarter ending December 2009 through the quarter ending December 2010, increasing from about 8,300 FTEs to 15,400 FTEs over that period. After December 2010, the number of FTEs began to decline because some recipients completed weatherization work as the end of the program draws near. As of the quarter ending September 30, 2011, FTEs declined to about 14,090 as weatherization work was completed and funds spent. DOE weatherization officials told us recipients are reducing the number of FTEs by an average of 200 jobs per month, which will continue to decrease until no FTEs remain by the scheduled end of the program in March 2012. Following Office of Management and Budget guidance, recipients reported on FTEs directly paid for with Recovery Act funding and not the employment impact on suppliers of materials (indirect jobs) or on the local communities (induced jobs). DOE weatherization officials and recipients told us that the number of jobs funded by the Recovery Act would be greater than the peak of 15,400 FTEs if the data on jobs funded included the associated jobs from the manufacturers of weatherization products, such as companies producing caulking guns or blower door technologies. Figure 7 shows the changes in the number of FTEs recipients reported from the quarter ending December 31, 2009, through September 30, 2011. According to DOE officials, all weatherization recipients reported to FederalReporting.gov for the quarter ending September 2011.

As in previous quarters, DOE performed a number of quality assurance steps on the data to help ensure the quality of the weatherization program's Recovery Act data. To support recipients' data quality, DOE asks the 58 recipients to report data on expenditures and weatherization production each month in DOE's PAGE system. DOE program officers who work directly with each recipient review the data submitted to ensure consistency. Also, DOE weatherization officials reported that they check data quality for data submitted by the 58 recipients in each recipient report against data collected in internal databases, such as PAGE. According to officials, data that do not correspond to the recipient report are flagged for comment and review. If discrepancies are found, DOE project officers work with the weatherization recipients to resolve them.

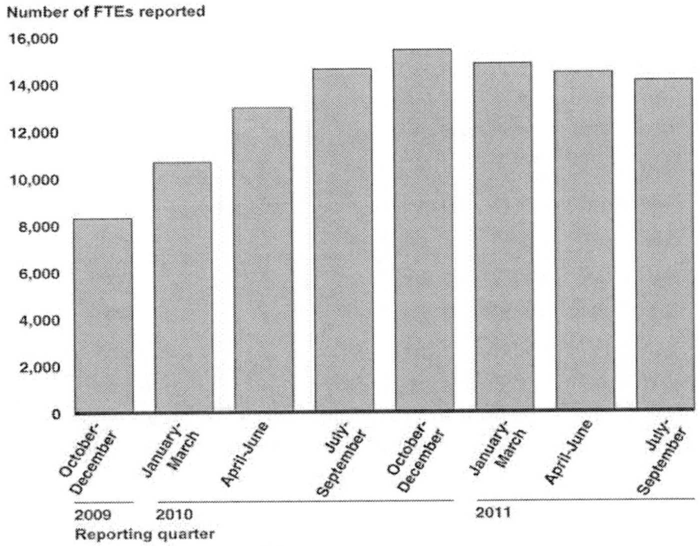

Source: GAO analysis of Recovery.gov data.

Note: Beginning in the quarter ending September 2010, FTEs include the Recovery Act weatherization program and the FTEs associated with the 34 training centers funded by the Recovery Act. The number of FTEs associated with the training center funds is minimal compared with the number of FTEs from the 58 recipients for the overall weatherization program. We did not include FTE data for the quarter ending September 2009 because of concerns about comparability with subsequent quarters of FTE reporting.

Figure 7. FTEs Recipients Reported for the Weatherization Assistance Program under the Recovery Act, October 2009 through September 2011.

According to DOE weatherization officials, DOE does not plan to use recipient-reported data internally, but the officials cited the data's usefulness in providing independent data on the numbers of jobs funded by the weatherization program. DOE weatherization officials told us that these data help support the program and its relevance by highlighting the extent to which DOE has achieved its program targets of funds spent, homes weatherized, and jobs funded.

AGENCY COMMENTS AND OUR EVALUATION

We provided a draft of this report to the Department of Energy for its review and comment. In its comments, DOE generally concurred with the majority of the findings and conclusions in our draft report. DOE noted that we characterized grant extensions differently on the first page than we did in the main body of the report; we added clarifying language in the highlights page to address DOE's comment. DOE also noted that, as of December 8, 2011, the number of recipients who have not received access to the additional Recovery Act funds is now three. We have added language to reflect the new number. Also in its comments, DOE provided clarifications to actions it has taken to address recommendations we made in May 2010 on the weatherization program under the Recovery Act. With respect to our May 2010 recommendation that DOE develop and clarify weatherization program guidance that clarifies the specific methodology for calculating the average cost per home, DOE stated that its September 2011 guidance provides a formula for this calculation. We concur and will close this recommendation as implemented. Regarding our May 2010 recommendation that DOE set time frames for states to develop and implement state monitoring programs, DOE identified training events and the issuance of a program notice announcing the rollout of "*The Weatherization Toolkit*," a compact disc that explains the steps to be followed in procuring materials and contractors. This recommendation remains largely unaddressed because these actions do not set time frames. Because we noted uneven progress among recipients in developing and implementing their monitoring programs, we believe that it is important for DOE to establish time frames. Moreover, the program notice cited by DOE predates our May 2010 recommendation, and we did not find it adequate for satisfying the need for time frames. DOE also provided technical comments, which we incorporated as appropriate.

Frank Rusco
Director, Natural Resources and Environment

APPENDIX I. OBJECTIVES, SCOPE, AND METHODOLOGY

Our objectives were to examine (1) the status and use of weatherization grant program funds under the Recovery Act; (2) the challenges, if any, that recipients faced in implementing the weatherization program under the Recovery Act; (3) the extent to which the weatherization program under the Recovery Act has achieved its energy and cost savings goals; and (4) the changes, if any, over time in the quality of full-time equivalent (FTE) data reported by Recovery Act recipients, particularly by weatherization program recipients.

To address these objectives, we reviewed relevant federal laws and regulations and the Department of Energy's (DOE) program notices on the use of funds, monitoring, reporting, and closeout requirements under the Recovery Act's weatherization program. We analyzed DOE data on production, expenditures, and projections of funds remaining.

We interviewed DOE weatherization officials who oversee the program about recipient progress in achieving production targets and spending their allocated funds, the grant closeout process, the program after the end of the Recovery Act, lessons learned, a potential deadline extension, the reliability of recipient reports, DOE's monitoring efforts, and progress on implementing GAO's previous recommendations.[63]

We also interviewed one project officer in DOE's field office in Golden, Colorado, and two at the National Energy Technology Laboratory; these project officers oversee several states in which we conducted site visits.

We met with officials from the National Association for State Community Services Programs, National Association of State Energy Officials, National Governors Association, and National Community Action Foundation to understand their role in helping their members implement the weatherization program under the Recovery Act and the challenges their members faced in using funds, monitoring projects, and reporting outcomes.

In September 2011 we attended the National Association for State Community Services Programs' Annual Conference in Seattle, Washington, where we met with a number of recipients and DOE weatherization officials.

To examine the status and use of weatherization grant program funds under the Recovery Act, we analyzed DOE-funding and production data provided by DOE.

We assessed the reliability of the data by interviewing knowledgeable DOE officials about the quality and potential limitations of the data and what checks and controls were in place to ensure data accuracy. We determined the data were sufficiently reliable for our purposes. We also selected a nonprobability sample of 10 states, which represents 17 percent of the 58 recipients of weatherization funding under the Recovery Act.

We selected these states using six criteria: the percentage of funds spent and homes weatherized, the amount of Recovery Act funding, geographic region, percentage of funding received out of the total Recovery Act weatherization funding, the average funds spent per completed household, and survey responses from the 55 of the 58 recipients to our web-based survey. In all 10 states, we interviewed state program representatives.

We conducted these interviews by telephone with state officials in Nebraska and Puerto Rico, and we visited state officials in Georgia, Maine, Maryland, Massachusetts, Oregon, Rhode Island, South Carolina, and Washington.[64]

In addition, in seven of these states,[65] we spoke with officials at a local agency,[66] and visited 11 Recovery Act-funded weatherization projects, including three training centers in 3 of the states; and one Sustainable Energy Resources for Consumer Grants project in Massachusetts. For a copy of the survey instrument, see www.gao.gov/ special.pubs/gao-12-195/index.html.

To examine the challenges, if any, that DOE and recipients faced in implementing the weatherization program under the Recovery Act, we first conducted a web-based survey of all 58 recipients; the survey went to the 58 state weatherization program managers and was open for 1 month. During that time, we also provided e-mail and telephone reminders.

We administered the survey to all 58 Recovery Act grant recipients and received responses from 55 recipients. The grant recipients who did respond may not represent all Recovery Act weatherization grant recipients.

Consequently, the survey results may not be generalized to include the Navajo Nation, the Northern Arapaho, or the District of Columbia—the recipients who did not respond to the survey. This survey asked recipients about revenue sources; the level of contact with project officers; the timeliness and helpfulness of DOE guidance; any challenges recipients faced in using funds, monitoring, and achieving program goals in the first 2 years of the Recovery Act; and any remaining challenges they face. See appendix III for a detailed explanation of the analysis of the survey data.

To further examine the challenges state program offices had implementing and monitoring their weatherization assistance programs under the Recovery Act, we also analyzed 15 DOE Inspector General and 27 state auditors' reports published between October 2009 and August 2011.

In addition, with the DOE project officers we interviewed, we discussed the initial challenges that they faced implementing the Recovery Act requirements, resolving recipient questions on program guidance, and the differences among the recipients they oversee in implementing weatherization programs using Recovery Act funds.

With recipients and local agencies, we discussed completing the Recovery Act requirements and finding additional sources of revenue to help continue their programs after March 2012. With officials from the four national associations, we discussed initial implementation challenges their members faced.

To determine the extent to which the weatherization program is expected to achieve energy and cost savings goals, we reviewed Oak Ridge National Laboratory's (ORNL) 2010 cost-benefit estimates of expected costs and benefits of the Recovery Act weatherization program and its plan for a more detailed evaluation of impacts for homes weatherized during the years 2009 through 2011.[67]

ORNL provided an electronic spreadsheet containing the model and the data that it used to estimate expected costs and benefits of the program. To better understand the model and sources of data used, including the electronic spreadsheet, we conducted follow-up interviews with the principal ORNL analyst in charge of the study.

The recipient reporting section of this report responds to the Recovery Act's mandate that GAO comment on the estimates of jobs created or retained by direct recipients of Recovery Act funds.[68] For our review of the ninth submission of recipient reports, covering July 1, 2011, through September 30, 2011, we built on findings from our eight prior reviews of the reports, which covered February 2009 through June 30, 2011.

To understand how the quality of jobs data reported by weatherization program recipients has changed over time, we compared the nine quarters of recipient reporting data that were publicly available at Recovery.gov as of October 31, 2011; performed edit checks; and conducted other analyses on weatherization recipients' reports for the Recovery Act.[69] We also matched DOE-provided data from the PAGE system. As part of the matching process, we examined the reliability of recipient data in the PAGE system with the information recipients reported directly to FederalReporting.gov.

Our reliability assessment included interviewing weatherization program officials and state weatherization officials and conducting logic tests for key variables. Our matches showed a high degree of agreement between funding information from the PAGE system and the information recipients reported directly to FederalReporting.gov . However, we observed that for half of the recipients, the discrepancy between the FTE value shown on their quarterly reports and the FTE value derived from the hours they reported to the PAGE system for the same period exceeded 10 FTEs.

Our examination of these discrepancies and interviews with DOE program officials indicated that the values observed in the PAGE data were most likely erroneous. DOE program officials stated that they track expenditures, but not FTE data, in PAGE. In general, we consider the recipient data used in this report to be sufficiently reliable, with attribution to official sources for the purposes of providing background information and a general sense of the status of recipients' reporting on the weatherization program.

To update the status of open recommendations from previous bimonthly and recipient report reviews, we obtained information from agency officials on actions taken in response to recommendations. Our findings based on analyses of FTE data are limited to the Recovery Act weatherization program and time periods examined and are not generalizable to any other programs' FTE reporting.

We conducted this performance audit from December 2010 through December 2011 in accordance with generally accepted government auditing standards. Those standards require that we plan and perform the audit to obtain sufficient, appropriate evidence to provide a reasonable basis for our findings and conclusions based on our audit objectives. We believe that the evidence obtained provides a reasonable basis for our findings and conclusions based on our audit objectives.

Appendix III. Technical Appendix (Survey)

Survey Design, Distribution, and Results

This appendix discusses survey, design, distribution, and results of our web-based survey of the 58 state-level recipients of the Weatherization Assistance Program funds under the Recovery Act and presents the survey we used. For a copy of the survey instrument, see www.gao.gov/special.pubs/gao-12-195/index.html.

To determine the extent to which recipients of the weatherization program funded under the Recovery Act spent Recovery Act funds, met monitoring requirements, and achieved weatherization program goals, we conducted a web-based survey of weatherization program recipients in the 50 states, the

District of Columbia, 5 territories, and 2 tribes. We asked recipients to provide information on their experience with the program and the extent to which they found certain aspects of the program implementation challenging, from April 2009 through June 2011.

We conducted pretests on our draft questionnaire to determine whether the (1) questions were clear and unambiguous, (2) terminology was used correctly, (3) questionnaire did not place an undue burden on state-level officials, (4) information could be readily obtained, and (5) survey was comprehensive and unbiased. We chose five pretest sites to include recipients in locations across a wide geographic area and in various stages of meeting spending and production targets. We conducted all five pretests over the telephone. We made changes to the content and format of the questionnaire after each of the pretests, based on the feedback we received.

We developed and administered a web-based questionnaire accessible through a secure server. Recipients were notified that the questionnaire was available online and were given unique passwords and usernames on May 16, 2011. We sent follow-up e-mail messages on May 31, 2011, June 9, 2011, and June 13, 2011, to those who had not yet responded. We contacted all remaining nonrespondents by telephone, starting on May 24, 2011. The questionnaire was available online until June 17, 2011. Questionnaires were completed by 55 weatherization program recipients, for a response rate of 95 percent. We telephoned the remaining three recipients (the District of Columbia, the Northern Arapaho Tribe, and the Navajo Nation), but they declined to participate within the study's time period.

Because this was not a sample survey, it has no sampling errors. However, the practical difficulties of conducting any survey may introduce errors, commonly referred to as nonsampling errors. For example, difficulties in interpreting a particular question, sources of information available to respondents, entering data into a database, or analyzing data can introduce unwanted variability into the survey results. We took steps in developing the questionnaire, collecting the data, and analyzing them to minimize such nonsampling error. For example, social science survey specialists designed the questionnaire in collaboration with GAO staff who had subject matter expertise. Then, we pretested the draft questionnaire with a number of state-level officials to ensure that the questions were relevant, clearly stated, and easy to understand. When we analyzed the data, an independent analyst checked all computer programs. Because this was a web-based survey, respondents entered their answers directly into the electronic questionnaire, eliminating the need to key data into a database, minimizing error.

Table 4. Percentage of Recipients Responding That Certain Areas Were Challenging and Not Challenging in 2009, the First Year of the Recovery Act-Funded Weatherization Assistance Program, and during June 2011

Percentage of recipients responding					
Challenges identified in survey	Initially (first year)		During June 2011 (third year)		Number responding
	Challenging	Not challenging	Challenging	Not challenging	
Balancing the requirements to meet production targets and to ensure the work is done correctly	90.6	9.4	67.9	32.1	53
Meeting production targets	90.6	9.4	56.6	43.4	53
Having additional reporting requirements	90.4	9.6	75.0	25.0	52
Implementing new program requirements that were developed as program progressed	88.9	11.1	63.0	37.0	54
Implementing Davis-Bacon requirements	88.2	11.8	39.2	60.8	51
Balancing training and technical assistance requirements with production targets	88.2	11.8	62.7	37.3	51
Adjusting to changes in existing reporting requirements	86.5	13.5	44.2	55.8	52
Meeting federal reporting requirements (FederalReporting.gov and PAGE)	84.3	15.7	62.7	37.3	51
Meeting certification and training requirements	83.0	17.0	56.6	43.4	53
Lacking expertise weatherizing multifamily units	82.0	18.0	64.0	36.0	50
Having too few monitoring staff	80.8	19.2	40.4	59.6	52
Having insufficient number of administrative staff	75.5	24.5	39.6	60.4	53
Having additional reporting requirements for monitoring	74.5	25.5	47.1	52.9	51
Lacking expertise in weatherizing multifamily units	74.0	26.0	60.0	40.0	50
Conducting the required number of monitoring visits	73.1	26.9	42.3	57.7	52
Adjusting to changes in existing reporting requirements for monitoring	72.0	28.0	48.0	52.0	50
Ensuring consistency in monitoring results	71.2	28.8	50.0	50.0	52
Lacking expertise in monitoring multifamily units	71.1	28.9	46.7	53.3	45
Complying with historic preservation requirements	70.6	29.4	35.3	64.7	51
Lacking trained monitoring staff	68.6	31.4	25.5	74.5	51
State-level auditors lacking training or expertise	63.3	36.7	22.4	77.6	49
Lacking staff to appropriately handle external monitoring (DOE-Inspector General, state-level auditors, consultants)	63.3	36.7	57.1	42.9	49

Percentage of recipients responding					
Challenges identified in survey	Initially (first year)		During June 2011 (third year)		Number responding
	Challenging	Not challenging	Challenging	Not challenging	
Determining how to measure long-term energy savings	59.6	40.4	51.1	48.9	47
Meeting state's requirements to monitor subgrantees	58.3	41.7	29.2	70.8	48
Lacking infrastructure at state level to handle the large influx of money from the Recovery Act	56.6	43.4	35.8	64.2	53
Determining how to measure cost savings (i.e., dollar savings)	54.3	45.7	37.0	63.0	46
Planning for cliff effect (i.e., decline in funding after Recovery Act funding expires in March 2012)	34.0	66.0	89.4	10.6	47
Determining and documenting income eligibility	25.0	75.0	17.3	82.7	52
Determining methodology for calculating $6,500 maximum average cost per home	17.6	82.4	3.9	96.1	51

Source: GAO analysis of recipients' responses.

In the report we showed the percentage of recipients responding to the survey who reported that certain program areas were strongly challenging or somewhat of a challenge, as opposed to not challenging. In table 4 we show the percentages of recipients responding who reported each area as initially challenging and not challenging, and still challenging and not challenging at the time of our survey. In almost all cases, there was decline in the percentage of recipients reporting the implementation area as continuing to be a challenge. The only exception was in responding to the challenges of the "cliff effect," or the decline in funding after Recovery Act funding expires in March 2012. For that item, the percentage of respondents who reported that it was challenging increased from 34 percent to 89 percent. For all of the other challenges, the percentage of recipients responding that a program area was challenging decreased and the percentage reporting an area as not challenging increased. For example, implementing Davis-Bacon requirements was reported to be challenging by 88 percent of respondents initially, but by only 39 percent of respondents at the time of the survey, June 2011. Also, lacking trained monitoring staff was reported to be challenging by 69 percent of respondents initially, but by only 26 percent of respondents at the time of our survey.

An overall decline in the percentage of respondents that report a specific item to be challenging could result from a more considerable shift of respondents from challenging to not challenging than from not challenging to challenging or, alternatively, from a smaller shift from challenging to not challenging accompanied by no shift in the reverse direction.

As shown in table 5, all but one challenge (planning for the cliff effect) moves in the direction from challenging to not challenging. In every case, this is more pronounced than moving in the opposite direction—from not challenging to challenging—though for a number of items some respondents identified something as initially not challenging but changed to identify as challenging at the time of the survey. For 16 of the 29 items, however, all of the change moved from challenging to not challenging.

Table 5. Percentage of Recipients Responding That Switched from Challenging to Not Challenging and Not Challenging to Challenging

Challenges identified in survey	Percent		
	Reporting initially challenging	Switching from challenging to not challenging	Switching from not challenging to challenging
Balancing the requirements to meet production targets and to ensure the work is done correctly	90.6	25.0	0
Meeting production targets	90.6	37.5	0
Having additional reporting requirements	90.4	17.0	0
Implementing new program requirements that were developed as program progressed	88.9	31.3	16.7
Implementing Davis-Bacon requirements	88.2	55.6	0
Balancing training and technical assistance requirements with production targets	88.2	28.9	0
Adjusting to changes in existing reporting requirements	86.5	48.9	0
Meeting federal reporting requirements (FederalReporting.gov and PAGE)	84.3	25.6	0
Meeting certification and training requirements	83.0	31.8	0
Lacking expertise weatherizing multifamily units	82.0	22.0	0
Having too few monitoring staff	80.8	57.1	30.0
Having insufficient number of administrative staff	75.5	47.5	0
Having additional reporting requirements for monitoring	74.5	42.1	15.4
Lacking expertise in weatherizing multifamily units	74.0	24.3	15.4
Conducting the required number of monitoring visits	73.1	47.4	14.3

Challenges identified in survey	Percent		
	Reporting initially challenging	Switching from challenging to not challenging	Switching from not challenging to challenging
Adjusting to changes in existing reporting requirements for monitoring	72.0	33.3	0
Ensuring consistency in monitoring results	71.2	32.4	6.7
Lacking expertise in monitoring multifamily units	71.1	34.4	0
Complying with historic preservation requirements	70.6	55.6	13.3
Lacking trained monitoring staff	68.6	62.9	0
State-level auditors lacking training or expertise	63.3	64.5	0
Lacking staff to appropriately handle external monitoring (DOE Inspector General, state-level auditors, consultants)	63.3	19.4	16.7
Determining how to measure long-term energy savings	59.6	17.9	5.3
Meeting state's requirements to monitor subgrantees	58.3	53.6	5.0
Lacking infrastructure at state level to handle the large influx of money from the Recovery Act	56.6	36.7	0
Determining how to measure cost savings (i.e., dollar savings)	54.3	32.0	0
Planning for "cliff effect" (i.e., decline in funding after Recovery Act funding expires in March 2012)	34.0	12.5	90.3
Determining and documenting income eligibility	25.0	38.5	26
Determining methodology for calculating $6,500 maximum average cost per home	17.6	88.9	2.4

Source: GAO analysis of recipient response.

APPENDIX IV. STATUS OF PRIOR OPEN RECOMMENDATIONS AND MATTERS FOR CONGRESSIONAL CONSIDERATION

In this appendix, we update the status of agencies' efforts to implement the 16 recommendations that remain open and are not implemented and 2 newly implemented recommendations that resulted from our Recovery Act mandate reports.[70] Recommendations that were listed as implemented or closed in a prior report are not repeated here. Last, we address the status of our matters for congressional consideration.

Table 6. Status of Prior Open Recommendations and Matters for Congressional Consideration

Department of Education	
Newly implemented recommendation	**Agency action**
1. To ensure all states receive appropriate communication and technical assistance for the State Fiscal Stabilization Fund (SFSF), consistent with what some states received in response to SFSF monitoring reviews, we recommend that the Secretary of Education establish mechanisms to improve the consistency of communicating monitoring feedback to states, such as establishing internal time frames for conveying information found during monitoring.[a]	1. Education agreed with our recommendation, and on November 21, 2011, submitted a letter to GAO discussing how it has refined its SFSF monitoring procedures to include time frames for communicating monitoring issues and feedback with states. For example, within five days of an exit meeting, the Department will send a state a written summary of the issues identified during the review and request any additional documentation to help resolve the issues. In addition, the Department will communicate in writing with the state at least once a month, if needed, for ongoing monitoring discussions; and twice a month to facilitate the resolution of monitoring issues after the issuance of the draft interim report.
Department of Energy	
Newly Implemented recommendation	**Agency action**
Weatherization Assistance Program	
1. Given the concerns we have raised about whether weatherization program requirements were being met, we recommended that DOE, in conjunction with both state and local weatherization agencies, develop and clarify weatherization program guidance that clarifies the specific methodology for calculating the average cost per home weatherized to ensure that the maximum average cost limit is applied as intended.[b]	1. In September 2011, DOE issued guidance that clarifies the methodology for calculating the average cost per unit weatherized and further defines the allowable cost categories used in calculating the average cost per unit.
Open recommendation	**Agency action**
Weatherization Assistance Program	
Given the concerns we have raised about whether weatherization program requirements were being met, we recommended that DOE, in conjunction with both state and local weatherization agencies, develop and clarify weatherization program guidance that:[b]	
1. Accelerates current DOE efforts to develop national standards for weatherization training, certification, and accreditation, efforts that are currently expected to take 2 years to complete.	1. DOE stated that it, along with the Department of Labor, released for comment a draft—"Workforce Guidelines for Home Energy Upgrades"—for single-family homes in November 2010. DOE weatherization officials noted that they received nearly 1,000 comments to the draft guidelines, which will delay publication of the final product.

Open recommendation	Agency action
2. Sets time frames for development and implementation of state monitoring programs.	2. As of October 2011, DOE had taken limited action to address the recommendation that it set time frames for the development and implementation of state monitoring programs. The recommendation remains largely unaddressed.
3. Revisits the various methodologies used in determining the weatherization work that should be performed based on the consideration of cost-effectiveness and develops standard methodologies that ensure that priority is given to the most cost-effective weatherization work. To validate any methodologies created, this effort should include the development of standards for accurately measuring the long-term energy savings resulting from weatherization work conducted.	3. In response to our recommendation that it revisit methodologies used to determine the most cost-effective weatherization work, DOE weatherization officials cited Oak Ridge National Laboratory's ongoing March 2014 evaluation. Officials told us the study results will be used to strengthen current protocols for determining weatherization work for both single and multifamily homes and to ensure cost-effectiveness.
4. Given that state and local agencies have felt pressure to meet a large increase in production targets while effectively meeting program requirements and have experienced some confusion over production targets, funding obligations, and associated consequences for not meeting production and funding goals, we recommended that DOE clarify its production targets, funding deadlines, and associated consequences while providing a balanced emphasis on the importance of meeting program requirements.	4. In response to our recommendation, DOE weatherization officials produced documentation showing extensive communication with the recipients over production and expenditures. However, they did not provide documentation showing how they clarified the consequences for not meeting these targets.
Energy Efficiency and Conservation Block Grant Program	
1. To better ensure that Energy Efficiency and Conservation Block Grant (EECBG) funds are used to meet Recovery Act and program goals, we recommended that DOE explore a means to capture information on the monitoring processes of all recipients to make certain that recipients have effective monitoring practices.[c]	1. DOE generally concurred with this recommendation, stating that "implementing the report's recommendations will help ensure that the Program continues to be well managed and executed." DOE also provided additional information on changes it has implemented. DOE added additional questions to the on-site monitoring checklists related to subrecipient monitoring to help ensure that subrecipients are in compliance with the terms and conditions of the award. These changes will help improve DOE's oversight of recipients, especially larger recipients, which are more likely to be visited by DOE project officers. However, not all recipients receive on-site visits. In November 2011, DOE also provided us with an update of its oversight of recipients following the addition of subrecipient monitoring questions to its on-site monitoring checklists. DOE officials stated that since July 2011, of the 171 recipients that have received a site visit by DOE staff, 162 were found to be using at least one of the monitoring practices identified by DOE.

Table 6. (Continued)

Open recommendation	Agency action
	In total, only 1 of the recipients visited was formally determined to be insufficiently meeting requirements for subrecipient monitoring. As noted previously, we continue to believe that the program could be more effectively monitored if DOE captured information on the monitoring practices of all recipients.
Department of Health and Human Services: Office of Head Start	
Open recommendation	**Agency action**
1. To help ensure that grantees report consistent enrollment figures, we recommended that the Director of the Department of Health and Human Services' (HHS) Office of Head Start (OHS) better communicate a consistent definition of "enrollment" to grantees for monthly and yearly reporting and begin verifying grantees' definition of "enrollment" during triennial reviews.[d]	1. OHS issued informal guidance on its website clarifying monthly reporting requirements to make them more consistent with annual enrollment reporting. This guidance directs grantees to include in enrollment counts all children and pregnant mothers who are enrolled and have received a specified minimum of services. According to officials, OHS is considering further regulatory clarification.
Department of Housing and Urban Development	
Open recommendation	**Agency action**
1. Because the absence of third-party investors reduces the amount of overall scrutiny Tax Credit Assistance Program (TCAP) projects would receive and the Department of Housing and Urban Development (HUD) is currently not aware of how many projects lacked third-party investors, we recommended that HUD develop a risk-based plan for its role in overseeing TCAP projects that recognizes the level of oversight provided by others.[e]	1. HUD responded to our recommendation by saying it must wait for project completion information to be entered into its reporting system in order to identify those projects that are in need of additional monitoring. HUD is currently in the process of identifying funding sources used for all TCAP projects completed as of September 30, 2011, and developing a report on that information. HUD will use the report to identify TCAP projects with little tax credit equity or no other federal funds leveraged that may need additional monitoring. HUD said it will develop a plan for monitoring those projects by November 30, 2011.
Office of Management and Budget (OMB)	
Open recommendation	**Agency action**
. We recommended that OMB provide timelier reporting on internal controls for Recovery Act programs for 2010 and beyond[f]	1. We previously reported that to address our recommendation, OMB had implemented a voluntary Single Audit Internal Control Project for states receiving Recovery Act funds.g One of the goals of the project was to achieve more timely communication of internal control deficiencies for higher-risk Recovery Act programs so that corrective action could be taken more quickly.

Open recommendation	Agency action
	The project encouraged participating auditors to identify and communicate deficiencies in internal control to program management 3 months sooner than the 9-month time frame required under statute. The project also required that program management provide corrective action plans aimed at correcting any deficiencies to the federal awarding agency 2 months earlier than required under statute. In Phase 1 of the project, which commenced in October 2009, 16 states participated and auditors audited 26 Recovery Act programs and reported 127 internal control deficiencies for which corrective action plans were developed by program managers. In Phase 2 of the project, which commenced in August 2010, 14 states partici-pated and auditors audited 36 individual Recovery Act programs and reported 173 internal control deficiencies for which corrective action plans were developed by program managers. In Phase 3 of the project, which commenced in October 2011, 10 states are participating and the project is expected to conclude in June 2012.
	While the project was successful in encouraging earlier reporting of internal control deficiencies for the participants, the voluntary nature of the project limited its effectiveness and may also have biased the project's results by excluding from analysis states or auditors with practices that cannot accommodate the project's requirement for early reporting of control deficiencies. We believe that OMB needs to take additional steps to encourage more comprehensive and timely reporting on internal control through single audits for Recovery Act programs. We will continue to monitor OMB's efforts for implementing this recommendation.
2. We recommended that OMB evaluate options for providing relief related to audit requirements for low-risk programs to balance new audit responsibilities associated with the Recovery Act.[h]	2. OMB has developed an option for providing some audit relief to the state auditors that participated in the Single Audit Internal Control project by modifying the requirements under OMB Circular No. A-133, *Audits of States, Local Governments, and Non-Profit Organizations*. The relief enabled the participating auditors to reduce the number of low-risk programs that were included in their risk assessment requirements. OMB officials acknowledged that as yet no viable alternative has been established to provide relief to all state auditors or other auditors that conduct single audits. However, according to OMB officials, initiatives are currently under way that, if approved, could provide audit relief in the future to auditors performing single audits. We will continue to monitor OMB's efforts in this area.

Table 6. (Continued)

Open recommendation	Agency action
3. We recommended that OMB issue Single Audit guidance in a timely manner so that auditors can efficiently plan their audit work.[i]	3, 4. OMB has worked with federal agencies and other entities involved in developing single audit guidance, and revised production schedules and deadlines in an effort to achieve an earlier issuance date. However, to date OMB has not yet achieved its goal of timely issuance of this important guidance. For example, OMB
4. We also recommended that OMB issue the OMB Circular No. A-133 Compliance Supplement no later than March 31 of each year.[j]	officials intended to issue the 2011 Compliance Supplement by March 31, 2011, but did not issue this guidance until June 1, 2011. OMB officials have developed a timeline for issuing the 2012 Compliance Supplement by March 31, 2012, and have begun working with federal agencies and others involved in issuing the Compliance Supplement. We will continue to monitor OMB's efforts in this area.
5. We recommended that OMB explore alternatives to help ensure that federal awarding agencies provide management decisions in a timely manner.[k] 6. We also recommended that OMB shorten the time frames required for issuing management decisions by federal agencies to grant recipients.[l]	5, 6. Regarding the need for agencies to provide timely management decisions, OMB officials identified alternatives for helping to ensure that federal awarding agencies provide management decisions for corrective action plans in a timely manner, such as shortening the time frames required for federal agencies to provide these decisions to grant recipients. OMB officials have acknowledged that this issue continues to be a challenge. In fiscal year 2011, most of the federal awarding agencies that had grantees with deficiencies identified as a result of the Single Audit Internal Control Project did not submit all of their management decisions for corrective actions by the specified due date. We will continue to monitor OMB's efforts in this area.
Department of Transportation	
Open recommendation	**Agency action**
1. To ensure that Congress and the public have accurate information on the extent to which the goals of the Recovery Act are being met, we recommended that the Secretary of Transportation direct the Department of Transportation's (DOT) Federal Highway Administration (FHWA) to develop additional rules and data checks in the Recovery Act Data System, so that these data will accurately identify contract milestones such as award dates and amounts, and provide guidance to states to revise existing contract data.[m]	1. In its response, DOT stated that it had implemented measures to further improve data quality in the Recovery Act Data System, including additional data quality checks, as well as providing states with additional training and guidance to improve the quality of data entered into the system. We have requested documentation of these changes, but DOT has not yet provided it. We are therefore keeping our recommendation on this matter open.

Open recommendation	Agency action
2. We recommended that the Secretary of Transportation direct FHWA to make publicly available—within 60 days after the September 30, 2010, obligation deadline—an accurate accounting and analysis of the extent to which states directed funds to economically distressed areas, including corrections to the data initially provided to Congress in December 2009.[n]	2. DOT stated that as part of its efforts to respond to our draft September 2010 report in which we made this recommendation on economically distressed areas, it completed a comprehensive review of projects in these areas, which it provided to GAO for that report. DOT recently posted an accounting of the extent to which states directed Recovery Act transportation funds to projects located in economically distressed areas on its website. We have requested documentation of the underlying data from DOT to verify this accounting, but DOT has not provided it. We are therefore keeping our recommendation on this matter open.
3. To better understand the impact of Recovery Act investments in transportation, we believe that the Secretary of Transportation should ensure that the results of these projects are assessed and a determination is made about whether these investments produced long-term benefits.[o] Specifically, in the near term, we recommended that the Secretary direct FHWA and Federal Transit Authority to determine the types of data and performance measures they would need to assess the impact of the Recovery Act and the specific authority they may need to collect data and report on these measures.	3. In its response, DOT noted that it expected to be able to report on Recovery Act outputs, such as the miles of road paved, bridges repaired, and transit vehicles purchased, but not on outcomes, such as reductions in travel time, nor did it commit to assessing whether transportation investments produced long-term benefits. DOT further explained that limitations in its data systems, coupled with the magnitude of Recovery Act funds relative to overall annual federal investment in transportation, would make assessing the benefits of Recovery Act funds difficult. DOT indicated that, with these limitations in mind, it is examining its existing data availability and, as necessary, would seek additional data collection authority from Congress if it became apparent that such authority was needed. Because DOT has not committed to assessing the long-term benefits of Recovery Act investments in transportation infrastructure, we are keeping our recommendation on this matter open.
Matters for Congressional Consideration—Office of Management and Budget	
Matter	**Status**
1. To the extent that appropriate adjustments to the Single Audit process are not the accomplished under current Single Audit structure, Congress should consider amending the Single Audit Act or enacting new legislation that provides for more timely internal control reporting, as well as audit coverage for smaller Recovery Act programs with high risk.[p]	1. We continue to believe that Congress should consider changes related to the Single Audit process.

Table 6. (Continued)

Open recommendation	Agency action
2. To the extent that additional coverage is needed to achieve accountability over Recovery Act programs, Congress should consider mechanisms to provide additional resources to support those charged with carrying out the Single Audit Act and related audits.[q]	2. We continue to believe that Congress should consider changes related to the Single Audit process.

Source: GAO.

[a] GAO-11-804, 47. [b] GAO-10-604, 124–125.

[c] GAO-11-379, 36. [d] GAO-11-166, 39.

[e] GAO-10-999, 189. [f] GAO-10-604, 247.

[g] GAO-10-999. [h] GAO-09-829, 127.

[i] GAO-10-604, 247. [j] GAO-10-999, 194.

[k] GAO-10-604, 247–248. [l] GAO-10-999, 194.

[m] GAO-10-999, 187-188. [n] GAO-10-999, 187-188.

[o] GAO-10-604, 241-242. [p] GAO-09-829, 128.

[q] GAO-09-829, 128.

End Notes

[1] Pub. L. No. 111-5, § 3, 123 Stat. 116 (2009).

[2] This amount was current as of October 14, 2011. For updated information, see Recovery.gov .

[3] Pub. L. No. 111-5§, 1512(e), 123 Stat. 115, 288. FTE data provide insight into the use and impact of the Recovery Act funds, but recipient reports cover only direct jobs funded by the Recovery Act. These reports do not include the employment impact on suppliers (indirect jobs) or on the local community (induced jobs). Both data reported by recipients and other macroeconomic data and methods are necessary to understand the overall employment effects of the Recovery Act.

[4] GAO, *Recovery Act: States' and Localities' Uses of Funds and Actions Needed to Address Implementation Challenges and Bolster Accountability,* GAO-10-604 (Washington, D.C., May 26, 2010).

[5] The grant recipients who did respond may not represent all Recovery Act weatherization grant recipients. Consequently, the survey results may not be generalized to include the Navajo Nation, Northern Arapaho, or the District of Columbia—the recipients who did not respond to the survey.

[6] We selected a nonprobability sample of 10 states, which represents 17 percent of the 58 recipients; because this is a nonprobability sample, information derived from this sample is not generalizable to all states but is illustrative. We spoke with officials in Georgia, Maine, Maryland, Massachusetts, Nebraska, Oregon, Puerto Rico, Rhode Island, South Carolina, and Washington.

[7] We conducted site visits to all the selected states, except Nebraska and Puerto Rico.

[8] In addition to conducting our analyses of recipient report data for the weatherization program under the Recovery Act, we continued, as in prior rounds, to perform edit checks and analyses on all prime recipient reports to assess data logic and consistency and identify unusual or atypical data.

[9] See http://www.gao.gov/recovery for related GAO products.

[10] "Homes" refers to housing units, which include single-family units, units within a multifamily building, and mobile homes. DOE defines a weatherized unit as a dwelling unit that has received a DOE-approved energy audit or been placed on a priority list and as a unit on which weatherization work has been completed and for which a final energy audit has taken place.

[11] DOE is required to allocate a base level of appropriated funds to states each year. According to DOE regulations, when the appropriated amount exceeds about $209 million, DOE distributes the remaining funds using an allocation formula. The allocation formula considers a number of factors but is particularly affected by the number of low-income households in a state. The grant awards for the two Indian tribes were subtracted from the overall grants to Arizona, New Mexico, and Wyoming.

[12] DOE originally set a target of weatherizing 593,000 low-income homes—that is, the total number of homes to be weatherized (production targets) as stated in state weatherization plans. As recipients amended their plans to reflect increased production, DOE revised its overall weatherization target to 607,000 homes.

[13] The Recovery Act required that DOE obligate funds for the program by September 2010; as a grant program, the Weatherization Assistance Program faced no spending deadline for recipients until DOE set one.

[14] Of the $450 million appropriation in fiscal year 2009, DOE received $200 million through its regular appropriations process. The remaining $250 million was supplemental funding appropriated by the Consolidated Security, Disaster Assistance, and Continuing Appropriations Act of 2009.

[15] Total LIHEAP appropriations in fiscal years 2009, 2010, and 2011 were about $5 billion, $5 billion, and $4.7 billion, respectively. The LIHEAP amounts available for weatherization

activities cited above are estimates based on a calculation of 15 percent of the total LIHEAP appropriations in each fiscal year. States may spend up to 15 percent of LIHEAP funds on weatherization activities, but each state agency administering LIHEAP determines the specific percentage of its LIHEAP funding to use on weatherization. In recent years, states have spent about 10 percent of their LIHEAP funds on weatherization, but state agencies may ask for a waiver in order to spend up to 25 percent. The Department of Health and Human Services submitted a request for $2.6 billion in LIHEAP funding for fiscal year 2012; of this total, 15 percent—about $390 million—would potentially be available for weatherization. Before the Recovery Act, the estimated amount available through LIHEAP for weatherization activities ranged from about $256 million to about $362 million.

[16] Weatherization program training and technical assistance funds support a range of activities, including measuring and documenting performance, monitoring programs, promoting advanced techniques, collaborating to further improve program effectiveness, training, and developing training tools and information resources.

[17] GAO-10-604.

[18] Furthermore, under Davis-Bacon, contractors and subcontractors are required to pay covered workers weekly and submit weekly certified payroll records to the contracting agency, generally the local agency.

[19] The deadline for completing DOE's Recovery Act funded weatherization training and technical assistance activities varies. Deadlines range from September 2011 through September 2013.

[20] Program operations and materials expenditures also include expenses related to health and safety, financial audits, liability insurance, leveraging, and special projects.

[21] GAO-10-604.

[22] GAO-10-604.

[23] GAO-10-604.

[24] DOE has released increments of the remaining 50 percent to some of the recipients who had not received full access to their weatherization funds as of September 2011. As of December 2011, DOE reported that three recipients still had not received access to additional Recovery Act funds.

[25] GAO-10-604.

[26] We received responses from 55 of the 58 recipients we surveyed, but not every respondent provided an answer to each question. Every question received a total of from 45 to 54 responses. See appendix III for more information on the survey. Also, the percentage reporting that an implementation action presented a challenge reflects responses identifying the action as a strong or somewhat of a challenge.

[27] We received 45 of 51 survey responses that identified implementing Davis-Bacon requirements as a strong challenge or somewhat of a challenge in the first program year.

[28] GAO-10-604.

[29] In July 2009, DOE and the Department of Labor issued a joint memorandum to weatherization program recipients, authorizing them to begin weatherizing homes using Recovery Act funds, with the condition to compensate workers for any differences in wage rates based on the final Department of Labor's determination of local prevailing wage rates for weatherization activities. On September 3, 2009, the Department of Labor completed its initial determination of wage rates for weatherization work conducted on residential housing units in each county. These wage rates were later revised in December 2009.

[30] We received 47 of 52 survey responses that identified having additional federal reporting requirements as a strong challenge or somewhat of a challenge in the first program year.

[31] Office of the State Auditor, North Carolina, *Single Audit Report for the Year Ended June 30, 2010* (Raleigh, N.C.: March 28, 2011).

[32] We received 36 of 50 survey responses identifying adjusting to changes in existing reporting requirements as a strong or somewhat of a challenge in the first program year.

[33] U.S. Department of Energy, Office of inspector General, *Preliminary Audit Report: Management Controls over the Commonwealth of Virginia's Efforts to Implement the*

American Recovery and Reinvestment Act Weatherization Assistance Program, OAS-RA-10-11 (Washington, D.C.: May 2010).

[34] We received 45 of 51 survey responses for the implementation area of balancing training and technical assistance requirements with weatherization production targets as a strong or somewhat of a challenge in the first program year.

[35] The Energy Independence and Security Act of 2007 amended the weatherization program definition of "state" to include the Commonwealth of Puerto Rico and the other territories and possessions of the United States. Consistent with the statutory requirement, in 2009 DOE amended the definition of "state" and amended the allocation procedure to include American Samoa, Guam, Northern Marianas Islands, Commonwealth of Puerto Rico, and the U.S. Virgin Islands.

[36] The funding data are based on the average amount of funding received by the state from 2004 through 2008. In fiscal year 2009, the weatherization program received regular appropriations that are separate from the Recovery Act funding.

[37] U.S. Department of Energy, Office of Inspector General, *Audit Report: Management Alert on the Department's Monitoring of the Weatherization Assistance Program in the State of Illinois,* OAS-RA-10-02 (Washington, D.C.: December 2009).

[38] State of Missouri, Missouri State Auditor, *Single Audit Year Ended June 30, 2010*, 2011-11 (Jefferson City, M.O.: March 2011).

[39] GAO-10-604.

[40] See appendix IV for more information on the status of GAO recommendations for the weatherization program.

[41] U.S. Department of Energy, Office of Inspector General, *Audit Report: The Department of Energy's Weatherization Assistance Program under the American Recovery and Reinvestment Act in the State of West Virginia,* OAS-RA-11-09 (Washington, D.C.: June 2011).

[42] New Jersey State Legislature, Office of the State Auditor, *Department of Community Affairs American Recovery and Reinvestment Act Weatherization Assistance Program Weatherization Agencies* (Trenton, N.J.: Nov. 8, 2010).

[43] GAO-10-604 .

[44] See appendix IV for more information on the status of GAO recommendations for the weatherization program.

[45] The website Weatherization Assistance Program Technical Assistance Center is a virtual library of all rules, regulations, policies, and procedures required by DOE's weatherization program. The site is maintained by the National Association of State Community Services Program, which is funded through a DOE grant.

[46] DOE's deadline applies not only to the distribution of funds by DOE and state-level agencies, but also to the use of the funds by the ultimate subrecipients to finish weatherizing homes.

[47] According to DOE's October 2011 guidance, any Recovery Act weatherization funds that remain unspent after closeout activities for the grant will be returned to the U.S. Treasury.

[48] DOE program officials originally told us that they had requested a 2-year extension, so that the new deadline would be March 2014. However the Office of Management and Budget released a memorandum in September 2011 stating that Recovery Act funds should be spent by September 2013.

[49] Fifty-two survey respondents answered the question about this implementation area. Of those 52, 47 respondents identified planning for an expected decline in funding after the end of Recovery Act funds as a strong or somewhat of a challenge in the third program year.

[50] The total amount remaining, about $317 million, is based on data as of August 2011. The data cover fiscal year 2008, fiscal year 2009, fiscal year 2010, and fiscal year 2011.

[51] Fifty-four of the 55 recipients responding to our survey answered the question about LIHEAP funds.

[52] The Regional Greenhouse Gas Initiative is a market-based regulatory program in 10 Northeastern and Mid-Atlantic states designed to reduce greenhouse gas emissions.

[53] The Electric Universal Service Program provides financial assistance with electric bills. The program is administered by the Maryland Office of Home Energy Programs.

[54] Oak Ridge National Laboratories, *Weatherization Assistance Program Technical Memorandum Background Data and Statistics,* prepared by Joel F. Eisenberg, March 2010 (ORNL/TM-2010/66).

[55] GAO reviewed an early draft of the ORNL evaluation plan, *Evaluation of the National Weatherization Assistance Program during Program Years 2009-2011 (American Reinvestment and Recovery Act Period),* prepared by Bruce Tonn, and others, May 20, 2011. (ORNL/TM-2011/87).

[56] For its estimates, ORNL considered the 50 states and the District of Columbia and not the Native American tribes and the U.S. territories that are also recipients of the weatherization program under the Recovery Act. ORNL assumed that the weatherization investment would yield energy savings over a 20-year period.

[57] The ORNL estimates did, however, account for differences in electric rates across regions of the United States. Hence, while making the simplifying assumption of a uniform reduction in electric consumption of 870 kilowatt-hours, the value of these savings varied across regions.

[58] In addition to increasing the number of workers employed directly in weatherizing homes, the weatherization program resulted in increased employment in the manufacture of materials and equipment used in weatherization and support services, according to ORNL.

[59] We limited our review of this document to the energy-only benefits discussion. We did not review the nonenergy benefits portion of the study plan.

[60] GAO-10-604.

[61] We also used our survey data from the 55 of the 58 weatherization recipients responding to our survey and interviews with selected recipients to identify the validation efforts recipients used.

[62] According to Recovery.gov, recipients reported on 58 grants across the weatherization program, in addition to 34 training center grants.

[63] In GAO, *Recovery Act: States' and Localities' Uses of Funds and Actions Needed to Address Implementation Challenges and Bolster Accountability,* GAO-10-604 (Washington, D.C., May 26, 2010), we made eight recommendations to DOE. See appendix IV for a detailed discussion of the status of the remaining five open recommendations.

[64] In Massachusetts, we also spoke with representatives from the Massachusetts Recovery and Reinvestment Office.

[65] In Oregon, we spoke with state officials but did not speak with local agencies or visit Recovery Act-funded projects.

[66] The local agencies we spoke with in 6 states were selected by the state program offices.

[67] The estimates come from Oak Ridge National Laboratory, W*eatherization Assistance Program Technical Memorandum: Background Data and Statistics,* prepared by Joel F. Eisenberg, March 2010 (ORNL/TM-2010/66). See the plan, Oak Ridge National Laboratory, *Evaluation of the National Weatherization Assistance Program during Program Years 2009—2011 (American Reinvestment and Recovery Act Period),* prepared by Bruce Tonn, et al., May 20, 2011 (ORNL/TM-2011/87).

[68] Pub. L. No. 111-5, § 1512(e), 123 Stat. 288.

[69] As with our previous reviews, we conducted these checks and analyses on all prime recipient reports to assess data logic and consistency and identify unusual or atypical data. For this ninth round of reporting, we continued to see only minor variations in the number or percentage of reports appearing atypical or showing some form of data discrepancy.

[70] GAO, *Recovery Act: As Initial Implementation Unfolds in States and Localities, Continued Attention to Accountability Issues Is Essential*, GAO-09-580 (Washington, D.C.: Apr. 23, 2009); *Recovery Act: States' and Localities' Current and Planned Uses of Funds While Facing Fiscal Stresses*, GAO-09-829 (Washington, D.C.: July 8, 2009); *Recovery Act: Funds Continue to Provide Fiscal Relief to States and Localities, While Accountability and*

Reporting Challenges Need to Be Fully Addressed, GAO-09-1016 (Washington, D.C.: Sept. 23, 2009); *Recovery Act: Recipient Reported Jobs Data Provide Some Insight into Use of Recovery Act Funding, but Data Quality and Reporting Issues Need Attention*, GAO-10-223 (Washington, D.C.: Nov. 19, 2009); *Recovery Act: Status of States' and Localities' Use of Funds and Efforts to Ensure Accountability*, GAO-10-231 (Washington, D.C.: Dec. 10, 2009); *Recovery Act: One Year Later, States' and Localities' Uses of Funds and Opportunities to Strengthen Accountability*, GAO-10-437 (Washington, D.C.: Mar. 3, 2010); *Recovery Act: States' and Localities' Uses of Funds and Actions Needed to Address Implementation Challenges and Bolster Accountability*, GAO-10-604 (Washington, D.C.: May 26, 2010); *Recovery Act: Opportunities to Improve Management and Strengthen Accountability over States' and Localities' Uses of Funds*, GAO-10-999 (Washington, D.C.: Sept. 20, 2010); *Recovery Act: Head Start Grantees Expand Services, but More Consistent Communication Could Improve Accountability and Decisions about Spending*, GAO-11-166 (Washington, D.C.: Dec. 15, 2010); *Recovery Act: Energy Efficiency and Conservation Block Grant Recipients Face Challenges Meeting Legislative and Program Goals and Requirements*, GAO-11-379 (Washington, D.C.: Apr. 7, 2011); *Recovery Act: Funding Used for Transportation Infrastructure Projects, but Some Requirements Proved Challenging,* GAO-11-600 (Washington, D.C.: June 29, 2011); *Recovery Act: Funds Supported Many Water Projects, and Federal and State Monitoring Shows Few Compliance Problems,* GAO-11-608 (Washington, D.C.: June 29, 2011); and *Recovery Act Education Programs: Funding Retained Teachers, but Education Could More Consistently Communicate Stabilization Monitoring Issues*, GAO-11-804 (Washington, D.C.: Sept. 22, 2011).

In: The Weatherization Assistance Program ISBN 978-1-62081-745-2
Editors: S. L. King and D. R. Griffiths

Chapter 3

WEATHERIZATION ASSISTANCE PROGRAM TECHNICAL MEMORANDUM: BACKGROUND DATA AND STATISTICS*

Joel F. Eisenberg

SUMMARY DATA SHEET

Households eligible for the Low Income Home Energy Assistance Program (LIHEAP)	38.6 million
Average site annual energy consumption for LIHEAP-Eligible Households	84 MBtu
LIHEAP-Eligible households annual energy consumption per square foot	90.2 Btu
Average LIHEAP-Eligible household energy expenditure for FY 2010	$1,800
Average first-year Weatherization energy savings	29 MBtu
Average first-year heating and cooling bill reduction (2010 prices)	$436.64
Range of potential annual base-load savings	$104–$174
Average annual CO2 reduction per weatherized home	2.65 metric tons
Benefit/cost ratio for energy savings	1.8
Societal benefit/cost ratio	2.5

* This is an edited, reformatted and augmented version of the Oak Ridge National Laboratory, dated March 2010.

INTRODUCTION

This technical memorandum is intended to provide readers with information that may be useful in understanding the purposes, performance, and outcomes of the Department of Energy's (DOE's) Weatherization Assistance Program (Weatherization). Weatherization has been in operation for over thirty years and is the nation's largest single residential energy efficiency program. Its primary purpose, established by law, is

> "…to increase the energy efficiency of dwellings owned or occupied by low-income persons, reduce their total residential energy expenditures, and improve their health and safety, especially low-income persons who are particularly vulnerable such as the elderly, the handicapped, and children."[1]

The American Reinvestment and Recovery Act PL111-5 (ARRA), passed and signed into law in February 2009, committed $5 Billion over two years to an expanded Weatherization Assistance Program. This has created substantial interest in the program, the population it serves, the energy and cost savings it produces, and its cost-effectiveness. This memorandum is intended to address the need for this kind of information.

Statistically valid answers to many of the questions surrounding Weatherization and its performance require comprehensive evaluation of the program. DOE is undertaking precisely this kind of independent evaluation in order to ascertain program effectiveness and to improve its performance. Results of this evaluation effort will begin to emerge in late 2010 and 2011, but they require substantial time and effort.

In the meantime, the data and statistics in this memorandum can provide reasonable and transparent estimates of key program characteristics. The memorandum is laid out in three sections. The first deals with some key characteristics describing low-income energy consumption and expenditures. The second section provides estimates of energy savings and energy bill reductions that the program can reasonably be presumed to be producing. The third section deals with estimates of program cost-effectiveness and societal impacts such as carbon reduction and reduced national energy consumption.

Each of the sections is brief, containing statistics, explanatory graphics and tables as appropriate, and short explanations of the statistics in order to place them in context for the reader. The companion appendices at the back of the memorandum explain the methods and sources used in developing the statistics.

SECTION 1. ENERGY-RELATED CHARACTERISTICS OF LOW-INCOME HOUSEHOLDS

Defining the Low-Income Population

There are a number of ways to define the households that can be described as low-income. Most of these are based on household income standards defined by various federal programs. For example, eligibility for many programs at the Department of Housing and Urban Development (HUD) is defined as household income at or below 80 percent of the local area median. The Department of Health and Human Services (HHS) Low Income Home Energy Assistance Program (LIHEAP) has historically defined eligibility to be household income at or below 150 percent of the Federal Poverty Income Guidelines or 60 percent of state median income, whichever is higher. The DOE Weatherization Assistance Program now defines eligibility as household income at or below 200 percent of the Poverty Income Guidelines or HHS LIHEAP eligibility. For purposes of this analysis, the LIHEAP guidelines are employed because they are consistent with the most reliable energy data regarding household energy use, the DOE Energy Information Administration (EIA) Residential Energy Consumption Survey (RECS)[2]. According to the latest RECS, there are 38.6 million households in the United States that are federally eligible for LIHEAP out of 111 million households nationwide.[3] Within this low-income population, 16.6 million households are categorized as having income below the poverty level. The low-income household population is generally distributed around the United States in roughly the same proportions as the non-low-income population, with approximately 21 percent in the Northeast, 24 percent in the Midwest, 36 percent in the South, and 19 percent in the West.

This large eligible population is composed of a broad range of households in terms of income levels, housing characteristics, and program participation. According to the 2005 RECS, more than 58 percent of the households contained at least one paid worker while approximately 21 percent received food stamps. Nearly 66 percent of the eligible households lived in single-family or mobile homes and more than half owned their own homes.

Fuel Types and Efficiency Characteristics

Low-income consumers have a similar profile to other consumers in terms of the primary heating fuel they use, with a slightly higher proportion than the general population using propane for this purpose. The largest single heating fuel type is natural gas, with approximately 48 percent of low-income households employing this fuel. Electricity is used for heating by 32 percent of these households. Home heating oil is the heating fuel of 8 percent of low-income families, and propane is employed for heating by 5 percent.

As the following graph demonstrates, the concentration of heating fuel usage varies substantially by region. Most of the home heating oil is used by households in the Northeast whereas electricity is the dominant heating source in the South, and natural gas predominates in the Midwest. These concentrations are found among both low-income and non-low-income households. This has important implications for the average level of residential expenditures and energy burdens in the various parts of the country because home heating oil and propane prices per million British thermal units (MBtu) are higher and have risen more sharply than those of other fuels. Furthermore, residential electricity prices in the northeastern part of the country also tend to be well above the national average. The overwhelming predominance of natural gas as the primary heating fuel in the Midwest makes low-income households in that part of the United States extremely sensitive to events in the natural gas markets.

Average annual weather-adjusted energy consumption for low-income households in 2005 was 84 MBtu compared to 102 MBtu for the non-low-income households. On the other hand, energy intensity, that is to say, Btu consumption per square foot of heated space, showed the reverse pattern. For eligible households, annual consumption per square foot averaged 90.2 Btu whereas for non-low-income households the average was 69 Btu per square foot. This reflects the relative inefficiency of the low-income housing stock compared to that of other households. Where 24 percent of low-income households reported inadequate insulation in their homes, 15 percent of non-low-income households reported this condition.

A review of the average MBtu consumption for low-income households by housing type reveals that households living in large apartment buildings have lower average annual consumption than those in most other building types (56.4 Mbtu compared to 99 Mbtu for those in single-family homes). It important to note that average consumption per square foot is far higher in the small multifamily housing stock than it is in other building types. The average consumption was 139 Btu per square foot for these homes compared to 81 Btu per square foot for single-family homes and 83 Btu per square foot for apartments in large buildings. This highlights a potential efficiency opportunity in the small multifamily housing stock.

By Region

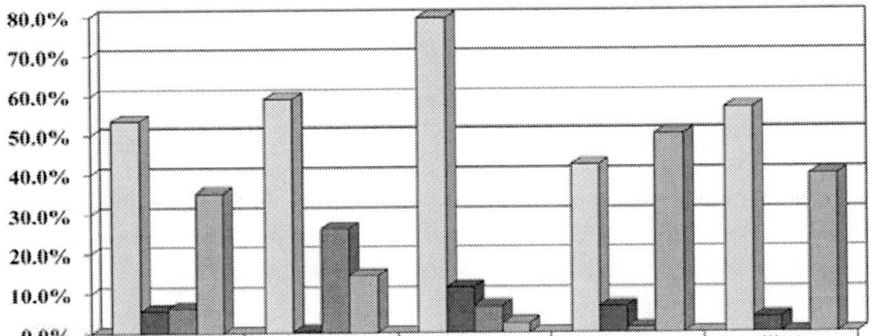

	U.S.	Northeast	Midwest	South	West
Natural Gas	53.4%	59.1%	79.6%	42.2%	56.7%
Propane	5.4%	0.4%	11.4%	6.4%	3.7%
Fuel Oil	6.0%	26.2%	6.5%	1.3%	0.0%
Electricity	35.2%	14.4%	2.4%	50.1%	39.6%

Source: DOE/EIA Residential Energy Consumption Survey for 2001

Primary heating fuel for Low-Income households.

Age of Heating Systems and Refrigerators, by Income Category

Age of Equipment	Refrigerators		Heating Systems	
	Low-Income	Non-Low-Income	Low-Income	Non-Low-Income
Under 2 Years	16.5%	17.4%	9.7%	11.4%
2–4 Years	19.3%	21.1%	10.9%	13.8%
5–9 Years	24.6%	29.5%	14.3%	21.7%
10–19 Years	22.9%	23.3%	19.2%	25.4%
20 Years or more	6.1%	5.2%	25.6%	19.7%
As old as home	10.6%	3.5%	20.3%	8.0%

The RECS data indicate substantial energy efficiency opportunities in the low-income housing stock in terms of both heating systems and refrigerators if one uses the age of the equipment as a rough proxy for inefficiency. Nearly 40 percent of the refrigerators in the low-income households were original installations or more than 10 years old compared to 32 percent for the non-low-income households. The contrast is comparable for heating systems, with 65 percent of low-income heating units being either original equipment or 10 or more years old. This compares to 53 percent for non-low-income households. Please see the table below for details.

Energy Expenditures and Burden

The estimated average annual expenditure for low-income households for FY 2010 (October 2009–September 2010) is $1,800, of which $747 (41.5%) is estimated to be for primary heating and cooling expenses. This compares to estimated residential energy expenditures of $2,231 for non-low-income households, of which $863 (38.7%) was for primary heating and cooling.

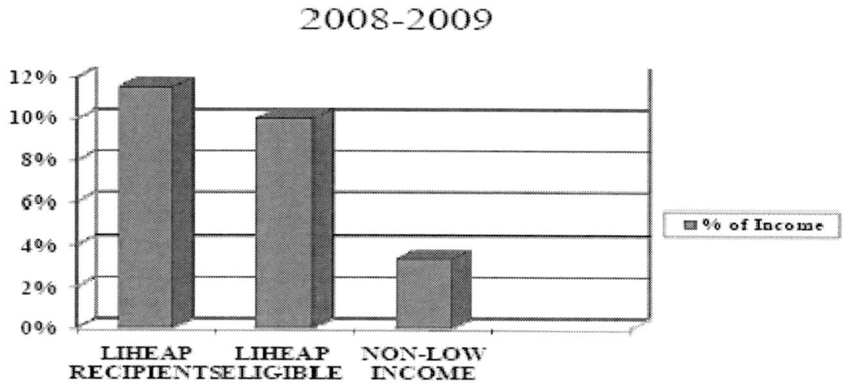

Source: Residential Energy Consumption Survey 2005

Energy Burden by Income Group.

Low-income households have lower average residential energy usage and lower residential energy bills than the non-low-income population, but this difference is not in proportion to household income. The average income of low-income households as provided in the 2005 RECS and adjusted for inflation was estimated at $18,624 compared to $71,144 for non-low-income households. In 2009 the group energy burden of low-income households, defined as average residential energy expense divided by average income, was estimated to be 10 percent of income for low-income households compared to

3.3 percent for non-low-income households as shown in the graph below. Households that actually received energy payment assistance, estimated at just over 5 million in 2005, had an even higher energy burden of 11.5 percent of income.[4]

As one examines the energy bills of eligible low-income households, several features stand out. Households that employ fuel oil as their primary heat source have the highest energy bills, followed by those that heat with propane. The average energy bill for the former was estimated at $3,194 for FY 2010 while that of the latter was $2,596. This compares with $1,684 for low-income households heating with natural gas and $1,569 for those heating with electricity. Please see the following graph for details of heating and total residential energy expenditures by primary heating fuel.

Given the concentration of low-income households heating with fuel oil in the Northeast it is no wonder that the average energy expenditure projected for low-income households for FY 2010 is higher there than elsewhere at $2,341. This compares to $1,745 for such households in the Midwest, where less-expensive natural gas predominates, and $1,770 for those in the South, where heating loads are lower. Those low-income households living in the West have a projected average expenditure of $1,328. The lower average expenditure in the West is caused primarily by lower average heating and cooling needs in that region.

SECTION 2. ENERGY SAVINGS AND BILL REDUCTION ESTIMATES

Estimated Energy Savings

Oak Ridge National Laboratory (ORNL) has employed engineering estimates derived from the Weatherization Assistant residential audit package to estimate annual energy savings resulting from the installation of cost-effective measures in low-income residential dwellings under the new parameters established for the Weatherization Program under the ARRA. The estimated annual savings per low-income household for heating and cooling are 29 MBtu. For comparison purposes, the most recent metaevaluation of Weatherization savings based on state-level evaluations using energy billing data was 30.5 MBtu (Schweitzer 2005).

The ARRA legislation modified the Weatherization Assistance Program in significant ways. Most important for purposes of this analysis, the average cost ceiling for investment per unit was set at $6,500, a substantial increase from the previously authorized ceiling of $3,000. Accordingly, the investment level per unit should increase, resulting in greater savings. However, the higher funding levels were accompanied by a shift in formula allocation, resulting in proportionally more funding flowing toward hot-climate states, which might act to reduce average savings. This study takes these factors into account, as explained in Appendix B, by running audits on typical low-income homes in each state, adjusting for housing types, and weighting for revenue allocation. Though the results below provide a reasonable examination of the issue of Weatherization energy savings and regional variations in them, it is important to keep in mind that input data on local weatherization costs and housing characteristics is presently limited and that the estimates below are exactly that, and not valid statistical representations of actual program performance.

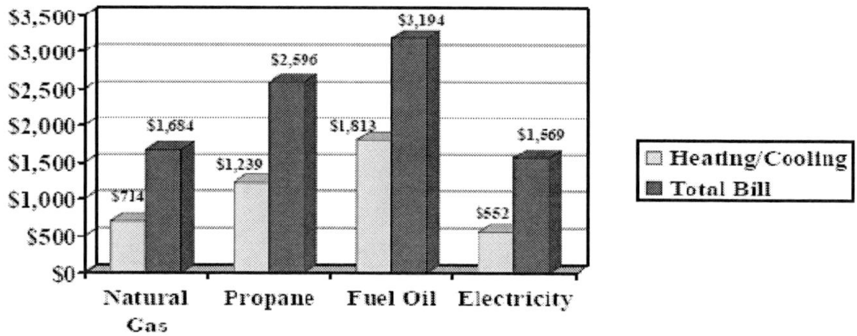

Source: ORNL Tabulation from EIA November 2009 STEO & 2005 RECS

Expected Energy Expenditures for 2009-2010 Low-Income Households by Primary Heating Fuel.

The weighted average national per-household expenditure savings, based on regional energy prices provided by EIA in its Annual Energy Outlook for 2010 and weighted by fuel use and housing type by state, is estimated at $436.64 in 2010. This figure represents the estimated annual savings only for heating and cooling measures. There is substantial uncertainty regarding savings resulting from base-load electric measures such as compact fluorescent bulb (CFL) change-outs and refrigerator replacements. A reasonable estimate for the savings resulting from those baseload measures is

$104 to $174 per household. This is an audit-based estimate; the relatively broad range reflects the uncertainty that comes from a lack of hard data on the extent to which these measures are applied.

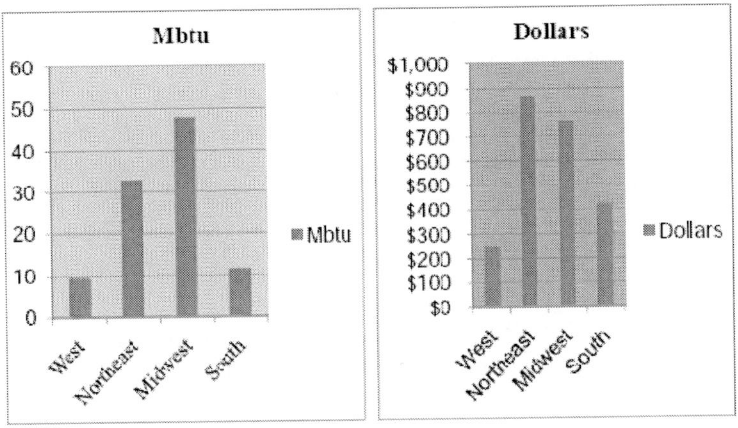

Sources: ORNL Weatherization Assistant Tabulations, EIA Annual Energy Outlook 2010.

Variations in Savings Selected States.

These data mask significant variations in both energy savings and expenditure reductions based on weather conditions, housing type, and energy costs. For example one large state in the Northeast had estimated energy savings of 32.93 MBtu compared to 47.92 MBtu in a large Midwest state for comparable homes. The higher MBtu savings were largely driven by differences in the weather conditions in the two locations. The estimated dollar savings, however, were higher in the Northeast state by more than $100 per year because of the higher energy prices in that location. Similarly, while two states in the South and West had comparable energy savings, the dollar savings per year in the South were estimated to be higher because of the higher energy prices in that particular location. Please see Chart for details.

SECTION 3. BENEFIT/COST RATIOS

Traditionally, the benefit/cost ratio for Weatherization has been expressed in terms of estimated national energy cost savings per dollar of investment for homes heated with natural gas, with cost savings based on national average residential natural gas prices. The estimates provided below reflect a somewhat different approach because they are based on state-by-state energy savings and investment numbers used to generate the savings estimates given in the previous section. As a consequence, the dollar savings estimates vary by region, by fuel type, by housing type, and with the adjusted average cost ceiling provided under ARRA. The estimated non-energy benefits are derived using the same methodology described above and, once again, the societal benefit/cost ratio is the combination of the energy and non-energy benefits divided by the estimated weighted average cost. Unlike the cost estimates that were derived from the metaevaluation, these estimates contain no assumed leveraged funding. The benefit/cost ratio estimated for energy benefits alone is 1.80 with total costs estimated at $5,704 and discounted energy benefits estimated at $10,253 in 2008 dollars. Non-energy benefits are estimated at $4,082 per household, also in 2008 dollars. The combined total benefit is estimated at $14,335, and the estimated societal benefit/cost ratio (all benefits divided by all costs) is 2.51. This is almost identical to the societal benefit/cost ratio of 2.53 computed in the most recent metaevaluation (Schweitzer 2005).

Estimated CO_2 savings, measured in metric tons per year of annual reduction, is estimated at 2.65 metric tons per unit weatherized and 53.1 metric tons over the life of the measures.

It is interesting to note that the estimated weighted average annual energy savings per household weatherized is slightly lower under this estimation methodology (30 MBtu) than the estimate in the previous metaevaluation (30.5 MBtu), although the dollar energy savings are substantially higher ($437 vs about $350). The former is partly the result of the fact that the new savings estimates are weighted for housing type variations under this method whereas the previous method provided estimates only for single-family homes, as in the metaevaluation. Perhaps even more important is the fact that the savings estimate in the metaevaluation was based on an assumed preweatherization energy consumption of 133 MBtu, which is much greater than the 84 MBtu found in this study. The dollar energy savings in this study, on the other hand, are substantially higher than those reported in the metaevaluation for two reasons. First, there is an attempt in this study to estimate energy bill reductions from the implementation of base-load electric measures such as CFLs and refrigerator replacement. For purposes of this effort, these base-load savings are estimated at 870 KWh per household. Second, by taking fuel type and cost into account (varied by region), the estimation methodology captures high fuel prices and expenditures in parts of the country that receive a substantial amount of weatherization funding and have relatively high fuel prices (e.g., the Northeast). This was not possible using the more simple methodology that was previously employed.

The methods used here are more complex and hopefully better at reflecting current reality than those used in previous years. Nonetheless, one needs to keep in mind the limitations of available input data that constrain an analysis such as this. There is no nationally available data on energy-related housing characteristics of weatherized households, nor is there data on the variations in the cost of measures installed from one locale to another. There is also no way of knowing at this point the exact characteristics of the housing stock and housing types that will actually get weatherized with the greatly expanded revenues available. Making these estimates more precise requires populating the methodology with more accurate data that will flow from the National Evaluation effort.

APPENDIX A. METHOD FOR ESTIMATING LOW-INCOME ENERGY EXPENDITURES

The method used to estimate the impact of projected price increases for residential energy on low-income households is based on the integration of two products from the Department of Energy's Energy Information Administration (EIA)—the Residential Energy Consumption Survey (RECS) for 2005 and the Short Term Energy Outlook (STEO) for November 2009. The former is the most recent EIA survey of U.S. households in an occasional series dating back to 1978 that provides detailed data on housing and energy characteristics, demographics, and energy consumption and expenditures verified through billing data. There are records on 4,382 individual household records in the data base. The STEO provides EIA's monthly estimate of energy prices in the 18-month period immediately ahead.

The RECS public-use files identify the location of each household by census region, of which there are four, and by census division, of which there are nine. Heating and cooling degree days are provided for each household for 2005 based on the population-weighted data for each census division; the poverty status of each household is also provided. The survey provides actual fuel bills for each household and uses statistical techniques to allocate the usage and expenditures among major usage categories such as heating, cooling, hot water heating, and refrigeration. Long-range climate normals for heating and cooling degree days for each of the census divisions were calculated using statistics provided by the National Climatic Data Center of the National Oceanic and Atmospheric Administration, National Environmental Satellite, Data and Information Service. These data were employed to calculate an adjustment factor for each division so that RECS data on heating and cooling expenditures for 2005 could be adjusted to reflect normal conditions.

Price adjustment factors were similarly calculated using quarterly price projections by division for natural gas and electricity and by region for propane and heating oil as provided by the STEO for November, 2009. The quarterly prices were weighted by consumption for each quarter to calculate an annual price adjustment factor for the historical record for 2009 and for the price projections for 2010. The baseline energy prices used to calculate multipliers were derived from the historical STEO database. This methodology provides a conservative and internally consistent approach to the estimation problem.

The estimate of an individual household's expenditure for a given year (n) was then calculated using the following formula in SPSS:

(((Dolngsph * hddfact) + Dolngwth + Dolngapl) * pmngd) + (((Dollpsph * hddfact) + Dollpwth + Dollpapl)*pmlp r) + (((Dolkrsph*hddfact + Dolkrwth+Doldrapl)*pmfo r) + (((Dolfosph*hddfact)+ Dolfowth + Dolfoapl)*pmfo r) + (((Dolelsph * hddfact) + (Dolelcol * cddfact) + dolelapl+ dolelrfg + dolelwth)* pmel d)

where

'Dol' signifies the expenditure for 2001,
'ng' represents natural gas
'lp' represents propane
'kr' represents kerosene
'fo' represents fuel oil
'el' represents electricity
'hddfact' is the heating degree day adjustment factor for normalization
'cddfact' is the cooling degree day adjustment factor for normalization
'sph' is space heating
'wth' is water heating
'apl' is appliances
'col' is cooling,
'rfg' is refrigerator
'pm' is price multiplier year 'n'

'r' is census region
'd' is census division.

Regional estimates were then made using the sort functions of SPSS to select households by region and qualification of eligibility for Low Income Home Energy Assistance Program and the 'Explore' statistical function to derive means, medians, and standard deviations by primary heating fuel type for total expenditures as well as heating and cooling expenditures for each year. Statistics were generated on a weather-normalized projected basis for FY 2010 (October 2009–September 2010).

APPENDIX B. METHODOLOGY FOR ESTIMATING ENERGY SAVINGS AND BILL REDUCTIONS

Certain metrics have been used in recent years to express the performance of the Weatherization Assistance Program. These include annual energy savings per household weatherized, expressed in million British thermal units (MBtu), first-year annual energy expenditure reductions expressed in dollars, the ratio of household life-time expenditure benefits to the cost of weatherization expressed as a benefit/cost ratio, and the ratio of total benefits, both energy and non-energy, to the cost of weatherization, expressed as the societal benefit cost ratio. In addition, the energy savings estimate has been used to derive estimated annual and lifetime CO2 reductions. These estimates have been based on the energy savings and cost estimates found in the most recent Oak Ridge National Laboratory (ORNL) metaevaluation of the Weatherization Assistance Program with prices and costs updated for inflation (Schweitzer 2005) and the ORNL analysis of non-energy benefits, also updated for inflation (Schweitzer and Tonn 2002).

The estimates in this memorandum are a methodological departure from this recently used procedure. There are several reasons that a change was needed. First, the metaevaluation estimates are dated and do not reflect recent changes in program operations that materially impact household savings. These include a major change in the program's average cost ceiling, from $3,000 to $6,500, an expansion in allowable measures to include electricity measures such as refrigerator replacement and lighting changeout, and a major increase in program funding impacting the allocation of resources among different regions and climate zones. Second, the metaevaluation results describe only homes heated with natural gas and do not reflect the diversity of heating fuels used in treated homes nor do they reflect potential cooling savings. The previous method reflected only savings in single-family homes and were never adjusted for variations in the treated housing stock. Finally, the previous estimates were based on national average energy prices and were not varied to reflect the diversity of energy prices weighted by the location of the weatherization work being performed.

The methodology used to prepare statistics for this memorandum and described below corrects for many of these deficiencies. Nonetheless, it does not represent a statistically valid representation of the Weatherization Assistance Program's performance and results. There are too many assumptions and uncertainties incorporated in it to allow that to be the case. Much of this is the result of a lack of up-to-date information on program operations, particularly regarding measures installed and their cost as well as the energy-related characteristics of the homes weatherized. Rather, the results should be treated as the best currently available estimate that can serve until more rigorous results are provided by the new National Evaluation.

The estimate is based on state-by-state projections of energy-efficiency investments, savings, and costs for a standard single-family home in moderate condition. Many states have very different weather conditions within their own borders, and in these cases, multiple assessments were made to take this into account, and an average result was derived. Census-division-level price projections were used to determine dollar savings and cost-effectiveness for individual measures. The prices were taken from the Energy Information Administration's Short Term Energy Outlook for September 2009.

The estimates were made using the Weatherization Assistant residential audit tool, version 8.4, and employing the measure installation costs contained in the Sample Supply Library provided by the audit. The home is a typical one-story wood-frame ranch home with 1,300 square feet of living space. There is some existing insulation in the attic but none in the wall cavities. The windows are moderately leaky and the foundation is an uninsulated slab. The air leakage rate as measured by a blower door test is 4,000 cfm at a house pressure difference of 50 pascals. The refrigerator is more than 15 years old and the heating system is in fair condition with a steady-state efficiency of 74 percent.

Information regarding the characteristics of actual homes weatherized is not available at the federal level, and this house cannot be confirmed as representative. For comparison purposes, the average heating and cooling consumption calculated for this house on a national basis is 83.3 MBtu. This compares to average low-income heating and cooling consumption of 47.1 MBtu for low-income homes in the 2005 Residential Energy Consumption Survey (RECS) and average pre-weatherization consumption for weatherized natural gas heated homes of 133 MBtu in the 1993 National Evaluation (Brown et.al. 2003).

The single-family energy-savings results were adjusted for the relative proportional consumption levels for manufactured homes and multifamily homes as provided in the RECS of 2005 and weighted for the proportion of each housing stock that was weatherized in a given state as reported to DOE for the 2008 Program Year. Primary heating fuel was adjusted to reflect regional fuel concentrations, such as home heating oil in New England and electricity in the Southeast. In the vast majority of cases, the average total of cost-effective investments in a home did not exceed the $6,500 average cost ceiling, but in the handful of cases where this was the case, the least cost-effective measures were eliminated until the ceiling was met.

There is a well-known history of actual realized results from weatherization activity yielding smaller savings than predicted by computerized audits. To take this effect into account the audit-predicted heating and cooling savings were adjust downward by 20 percent so as not exaggerate program performance. In addition, it was assumed that only 60 percent of the homes actually received a new refrigerator. There are no data indicating the proportion of weatherized homes that receive new refrigerators, so the 60 percent assumption was derived from a review of a recent evaluation of the Weatherization Assistance Program in Ohio.

The energy bill reductions were calculated by deriving a fuel mix for each state's weatherized homes as reported for the 2008 Program Year and deriving a weighted average price per MBtu using census-division-level prices from the Energy Information Administration's (EIA's) Annual Energy Outlook 2010 Early Release, (EIA AEO 2009).

A weighted average national cost for weatherization was derived by dividing the audit-generated state cost into the state's ARRA allocation in order to develop a total number of state units weatherized and dividing this figure into the sum of all weatherized units to develop a state weighting factor. This weighting factor was then used to develop weighted average national weatherization per-unit costs, energy savings, and bill reductions.

The benefit/cost ratios were calculated assuming a 20-year life of the measures as was the case in the original 1993 National Evaluation. Energy prices over the 20-year period were provided by the EIA AEO in 2008 constant dollars and savings were discounted at 3 percent per Office of Management and Budget (OMB) Circular. A-94 Appendix C (OMB 1992).

Carbon dioxide reductions were calculated for each state by weighting the carbon content of the heating and cooling energy saved by the carbon content of the fuels used in weatherized homes in that state. Reductions from base-load measures were calculated from a national average content of carbon per kilowatt-hour consumed. Total reductions were also calculated based on an assumed 20-year life of the measures. These reductions were calculated for each state and then weighted by the proportion of all homes weatherized as described earlier for the calculation of national average energy savings.

APPENDIX C. LIST OF REFERENCES

[1] Brown, M. et. al. (1993). *National Impact of the Weatherization Assistance Program in Single-Family and Small Multifamily Dwellings*. ORNL/CON-326. Oak Ridge National Laboratory, Oak Ridge, Tenn.

[2] Eisenberg, J. (2005). *The Impact of Forecasted Energy Price Increases On Low-Income Consumers*. ORNL/CON-495. Oak Ridge National Laboratory, Oak Ridge, Tenn.

[3] Energy Information Administration. (2009). *Annual Energy Outlook for 2010, Early Release*. Washington, D.C.

[4] Energy Information Administration,.(2008). *Residential Energy Consumption Survey for 2005*. Public Use Microsample, Washington, D.C.

[5] Energy Information Administration (2005). *Short Term Energy Outlook, November*. Washington, D.C.

[6] Office of Management and Budget (OMB). (1992). *Memorandum for Heads of Executive Departments and Establishments*. Circular. A-94. Washington, D.C.

[7] Schweitzer, M. (2005). *Estimating the National Effects of the U.S. Department of Energy's Weatherization Assistance Program with State-Level Data: A Metaevaluation Using Studies from 1993–2005*. ORNL/CON-492. Oak Ridge National Laboratory, Oak Ridge, Tenn.

[8] Schweitzer, M. & B. Tonn. (2002). *Nonenergy Benefits from the Weatherization Assistance Program: A Summary of Findings from the Recent Literature*. ORNL/CON-484. Oak Ridge National Laboratory, Oak Ridge, Tenn.

[9] U.S. Census Bureau. (2006). *Income, Poverty, and Health Insurance Coverage in the United States*. Washington, D.C.

[10] U.S. Department of Commerce, National Oceanic and Atmospheric Administration. (2007). *State, Regional, and National Heating Degree Days*. Historical Climatology Series 5–1, Monthly. , Asheville, N.C.

[11] U.S. Department of Health And Human Services, Administration for Children and Families. (2004–7). *LIHEAP Home Energy Notebook*, Washington, D.C.

[12] U.S. National Weather Service, Climate Prediction Center. (2007). *Experimental Monthly Total Degree Day Forecast*. Camp Springs, Md.

End Notes

[1] Title 42 of the U.S. Code, Chapter 81, Subchapter III, Part A, 6861.

[2] Energy expenditure statistics in this memorandum are based on the public use files of the 2005 RECS, weather-normalized, and adjusted for FY 2010 prices as projected by EIA.

[3] Special tabulations from the American Community Survey, 2005-2007, which has more precise income data, indicate that there were approximately 32.8 million households at or below 200 percent of the federal poverty income guidelines and 34.9 million households at or below 60 percent of state median income.

[4] There are two ways that energy burden is generally calculated. Mean Group Burden is computed by calculating the average energy expenditure for all households in a group of interest and dividing by the average income of households in that group. Mean Individual Burden is calculated by dividing each household's energy bill by its income and calculating an average of these individual burdens for the target group. The latter method produces a low-income energy burden of 14.2 percent compared to 10 percent for the group burden. The group method is used in this report because of uncertainties surrounding the broad income ranges in the 2005 RECS data, but both statistics are valid descriptions of low-income energy burden.

In: The Weatherization Assistance Program ISBN 978-1-62081-745-2
Editors: S. L. King and D. R. Griffiths

Chapter 4

PROGRESS IN IMPLEMENTING THE DEPARTMENT OF ENERGY'S WEATHERIZATION ASSISTANCE PROGRAM UNDER THE AMERICAN RECOVERY AND REINVESTMENT ACT*

United States Department of Energy

MEMORANDUM FOR THE SECRETARY

FROM: Gregory H. Friedman
Inspector General

SUBJECT: INFORMATION: Special Report on "Progress in Implementing the Department of Energy's Weatherization Assistance Program Under the American Recovery and Reinvestment Act

* This is an edited, reformatted and augmented version of the Special Report, Under the American Recovery and Reinvestment Act, U.S. Department of Energy, OAS-RA-10-04, dated February 19, 2010.

Background

The American Recovery and Reinvestment Act of 2009 (Recovery Act) was enacted to stimulate the U.S. economy, create jobs and make infrastructure investments in energy and other areas. The Department of Energy's (Department) Weatherization Assistance Program received about $5 billion under the Recovery Act to improve the energy efficiency of nearly 590,000 residences of low income citizens -- a dramatic increase over the $450 million appropriated for this Program in Fiscal Year 2009. The Department awarded $4.73 billion of the Recovery Act funding in the form of grants to all 50 states, 5 territories, the District of Columbia and 2 Native American Tribes (the remaining $270 million was retained by the Department to administer the grants). It was anticipated that the expenditure of these funds would result in the almost immediate creation of jobs because they were to rely on the existing Weatherization Program infrastructure, recognizing that many homes of low income citizens were in need of energy usage improvements. As of December 2009, grantees had been authorized to spend up to 50 percent of awarded funds, with additional monies to be provided based on performance. Corresponding to funding increases, weatherization production goals increased significantly, rising from 104,000 in 2009 to 586,015 units over the 3 year life of the Recovery Act. As noted in our *Special Report on the American Recovery and Reinvestment Act at the Department of Energy* (OAS-RA-09-01, March 2009), the Department of Energy's grant-making authority, an essential component of the Weatherization Program's strategy, was critical to achieving the desired Recovery Act-related economic stimulus. In that report, we suggested that the Department take steps to develop aggressive safeguards to ensure that performance was monitored throughout the life-cycle of the grants. Because of the importance of this program to stimulating the economy, creating jobs, and improving the quality of life in low income households, we initiated this review to provide the Department with an interim status report highlighting factors impacting progress in meeting Weatherization Assistance Program and Recovery Act goals.

Observations and Conclusion

The Department had taken a number of proactive steps to foster timely implementation of the Weatherization Program. However, in spite of the

Department's efforts, grantees had made little progress in weatherizing homes. As of February 2010, the one-year anniversary of the Recovery Act, only a small percentage of Recovery Act weatherization funds had been spent and few homes had actually been weatherized:

- Only \$368.2 million (less than 8 percent) of the total award of \$4.73 billion had been drawn by grantees for weatherization work (please see Appendix I); and,
- Corresponding to the low spending rates, grant recipients fell significantly short of goals to weatherize homes. Notably, as shown in the following chart, only 2 of the 10 highest funded recipients completed more than 2 percent of planned units. If the state of Ohio, one of the most productive jurisdictions, is taken out of the mix for comparative purposes, the completion rate for the remaining nine states was well under two percent.

State	Units Planned	Units Weatherized as of December 2009	Percentage of Units Completed
New York	45,400	280	0.62%
Texas	33,908	0	0.00%
Ohio	32,180	6,814	21.17%
Pennsylvania	29,554	378	1.28%
Michigan	33,410	385	1.15%
Illinois	26,933	331	1.23%
California	43,400	12	0.03%
Florida	19,090	312	1.63%
Wisconsin	20,678	772	3.73%
North Carolina	22,203	197	0.89%

Source: Department's National Weatherization ARRA Snapshot Report dated February 16, 2010.

Appendix II to this report provides Weatherization Program data for all 58 recipients. As noted in the Appendix, 13 of these recipients, including a number of large states, had not weatherized any residences as of February 16, 2010. This finding was based on data provided by the Department as of February 16, 2010.

Because the consequences of the lack of progress by grantees in the implementation of the Weatherization Program were so significant, we found this data alarming. In short, the Nation has not, to date, realized the potential economic benefits of the \$5 billion in Recovery Act funds allocated to the

Weatherization Program. The job creation impact of what was considered to be one of the Department's most "shovel ready" projects has not materialized. And, modest income home residents have not enjoyed the significant reductions in energy consumption and improved living conditions promised as part of the massive Recovery Act weatherization effort.

To its credit, the Department, through site visits to 32 states and numerous interactions with state officials, identified challenges facing the grantees that delayed weatherization progress. Department officials worked aggressively with the states and other responsible agencies to mitigate these challenges. Further, we were informed that most grantees were able to weatherize some residences during Fiscal Year 2009 using funds awarded prior to the enactment of the Recovery Act. However, as a practical matter, program challenges, such as those identified in this report, placed the Recovery Act-funded Weatherization Program "on hold" for up to nine months. Despite its best efforts, the Department's actions to reduce or eliminate program delays appeared not to have significantly increased the tempo of actual units weatherized across the Nation.

Regulatory Requirement

The Recovery Act required that recipients of weatherization funds pay laborers at least the prevailing wage as determined under the Davis-Bacon Act. This requirement was not previously applicable to weatherization activities, and as such, grantees lacked information on which to base wage rates. In response, the Department asked the Department of Labor (Labor) to provide necessary wage determinations for each of the geographical areas expected to receive weatherization funds. Labor then began to conduct wage surveys across the country to determine the appropriate wage for weatherization work.

Recognizing that delays would occur while the wage surveys were completed, Federal officials attempted to mitigate the effects of implementing the Davis-Bacon Act. The Secretaries of Energy and Labor issued a joint memorandum in July 2009 to grantees stating that state and local agencies should begin weatherizing homes "now" while they were waiting for the results of the surveys. The guidance recommended proceeding with weatherization efforts and adjusting salaries retroactively, if workers had been paid less than the appropriate wage rates. In many cases, however, states elected not to proceed with weatherization efforts as recommended. They were concerned with avoiding perceived administrative problems and burdens associated with retroactive adjustments to wages. Many grantees chose not to begin work until the prevailing wage rates were formally established. Even after Labor's work was complete, additional delays occurred while grantees prepared guidance for sub-recipients on how to apply the wage rates. As such, efforts to begin the work of weatherizing homes did not begin in earnest until after guidance was completed in October 2009.

State-Level Issues

A number of other factors at the state level also contributed to delays in developing and implementing plans for the Weatherization Program. Among other things, certain states were unable to prepare their required program plan and submit it to the Department for review and approval in a timely manner. Ironically, given the anticipated stimulus effect of the program, economic problems in many states adversely impacted their ability to ensure that weatherization activities were performed. State hiring freezes, problems with resolving significant local budget shortfalls, and state-wide planned furloughs delayed various aspects of the program and contributed to problems with meeting spending and home weatherization targets.

Several grantees from which we obtained information encountered other delays in preparing their Weatherization Program plans. The Department required all grantees to submit plans for implementing the Weatherization Program by May 2009. The state plans were to outline the use of Recovery Act funds and detail performance metrics for weatherizing homes. Upon approval of the plans by the Department, grantees were eligible to receive 40 percent of their Weatherization Program allocation. The statewide Weatherization Assistance Program in North Carolina, for instance, encountered difficulties in finalizing its plan when the administration of the program was transferred to the State's Department of Commerce in August. Although North Carolina's original state plan had been approved by the Department in June 2009, the State had to submit an amended plan, including budget information, through its Department of Commerce. The amended state plan was ultimately not approved by the Department of Energy until November 2009. Accordingly, North Carolina did not have access to its allocation for the Weatherization Program until that time. In another instance, the State of Wisconsin did not have an approved plan in place until September 2009. Subsequently, the State amended the plan with the new wage determinations. Until it had incorporated the new wage determinations issued in December 2009 into its plan, Wisconsin chose not to proceed with amendments to sub-recipient contracts, thus delaying weatherization production. Certain states also faced fiscal challenges that contributed to the delay in the implementation of the Weatherization Program. For example, because of budget shortfalls associated with the economic downturn, certain states were under hiring freezes that applied to all employees regardless of the source of their funding, including those tasked with weatherization-related work. In other states, progress was impacted because personnel involved with the program were subject to significant state-wide furloughs. Further, the approval of state budgets was delayed in states such as Pennsylvania as legislators deliberated over how to address overall budget shortfalls. Lacking staff, states were unable to perform required implementation tasks necessary to handle the large infusion of Recovery Act Weatherization Program funding. Without budgets, states did not have spending authority and, hence, were not able to obligate or expend any Weatherization Program funds. Officials in California, one of the largest recipients of weatherization funds, reported that furloughs created significant staffing challenges in implementing the Weatherization Program. Similarly, officials in Illinois indicated that their plan to hire staff necessary to implement their weatherization plans had been delayed because of a hiring freeze in the State. Illinois projected an additional need for 21 staff members to meet its Recovery Act goals. As we noted in our *Management Alert on the Department's Monitoring of the Weatherization Assistance Program in the State of Illinois* (OAS-RA-10-02, December 2009), the lack of staff adversely impacted its ability to implement its Weatherization Program plan. The hiring freeze in Illinois was lifted in November 2009 and a sustained hiring effort was underway. A hiring freeze in New York City also slowed the implementation of weatherization work in that jurisdiction.

Training

To ensure successful implementation of the Recovery Act requirements for the Weatherization Program, the Department required that program personnel at the state and community action agency level receive additional training. The training was required to ensure that recipients were familiar with requirements associated with the Davis-Bacon Act, new eligibility requirements for homes to be weatherized, increased allowable costs per unit, and monitoring of work performed by sub-recipients. As with other activities, necessary state training initiatives were delayed by budget shortfalls and staff furloughs. As a result, additional time was needed to develop effective training programs and to deliver the training to impacted personnel.

PATH FORWARD

During consideration of the Recovery Act and following its enactment, the conventional wisdom was that the Department's Weatherization Program was about as close to meeting the definition of "shovel ready" as virtually any program in the Department's portfolio. Specifically, the Recovery Act weatherization effort had the following attributes:

- An existing programmatic infrastructure, including processes and procedures which had been in place for many years;
- The techniques for weatherization tasks were well known and comparatively uncomplicated, and the requisite skills were widely available;
- Performance metrics were relatively easy to establish and understand;
- The potential benefits for low income citizens were easily recognized; and,
- The potential beneficial impact on energy conservation was obvious.

Under these circumstances, it seemed that the $5 billion in Recovery Act funds allocated to this program would have had a prompt and easily discernable impact on job creation and economic stimulation. These goals have proven to be much more difficult to achieve than originally envisioned.

The results of our review confirmed that as straight forward as the program may have seemed, and despite the best efforts of the Department, any program with so many moving parts was extraordinarily difficult to synchronize. In this case, program execution depended on the ability of the Federal government, (multiple agencies, in fact) state government, grant sub-recipients and weatherization contractors, working within the existing Federal and state regulatory guidelines, to respond to a rapid and overwhelming increase in funding. Further, anticipated stimulus impact was affected by certain conditions and events clearly outside of Departmental control including state budget difficulties; availability of trained and experienced program staff; and, meaningful changes in regulatory requirements. As these issues are resolved, it appears likely that pressure will increase to accelerate the weatherization of residences in the compressed statutory timeframe available under the Recovery Act. In a situation like this, our concern is that the understandable desire to spend the Weatherization funds on a catch-up basis may lead to an environment conducive to wasteful, inefficient and, perhaps even abusive practices. To its credit, the Department recognized the need to ensure adequate monitoring of the Weatherization Program, and is in the process of hiring staff for this purpose. However, given the increased risk associated with accelerating expenditures, we recommend that the Department re-evaluate its monitoring and staffing plans and adjust them as necessary to prevent, detect, and remedy instances of fraud, waste and abuse of Recovery Act funds.

MANAGEMENT COMMENTS

Management provided comments which expanded upon the findings presented in this report. The Department responded that it continues to ramp up to achieve the full impact of this important program. Management expressed its resolve to work diligently to achieve robust spending while maintaining proper programmatic oversight. Management also told us that Departmental officials have worked closely with states in incorporating new Weatherization Assistance Program Recovery Act requirements, such as the Davis-Bacon Act prevailing wage provisions. The Department also asserted that it is taking proactive steps to accelerate the program schedule, such as issuing 12 program guidance documents to provide grantees with clarification on important topics, including the implementation of the Davis-Bacon Act and monitoring requirements. We considered management's comments to be integral to any consideration of the current status of the Department's Weatherization efforts. For that reason, other than technical issues which we considered in the report process, management's comments are provided in their entirety at Appendix III of this report.

cc: Deputy Secretary
Under Secretary of Energy
Chief of Staff
Assistant Secretary for Energy Efficiency and Renewable Energy, EE-1
Director, Office of Risk Management, CF-1.2
Team Leader, Office of Risk Management, CF-1.2
Audit Liaison, EE-3A

APPENDIX I

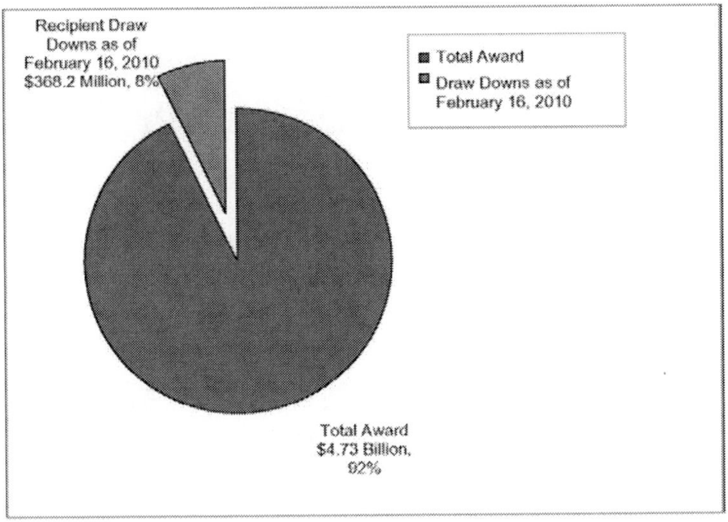

Source: Department's National Weatherization ARRA Snapshot Report dated February 16, 2010.

Distribution of Weatherization Recovery Act Funds to States and Territories.

APPENDIX II

Total Units Weatherized Compared to Total Units Planned As of February 16, 2010

Recipients	Total Completed Units	Total Units Planned	Percent of Units Completed
Alabama	477	6,651	7.17%
Alaska	0	1,523	0.00%
Arizona	359	6,409	5.60%
Arkansas	625	5,578	11.2%
California	12	43,400	0.03%
Colorado	1,369	10,478	13.07%
Connecticut	23	7,500	0.31%
Delaware	519	1,526	34.01%
District of Columbia	0	785	0.00%
Florida	312	19,090	1.63%
Georgia	632	13,871	4.56%

Recipients	Total Completed Units	Total Units Planned	Percent of Units Completed
Hawaii	0	672	0.00%
Idaho	551	3,113	17.70%
Illinois	331	26,933	1.23%
Indiana	974	19,736	4.94%
Iowa	370	7,196	5.14%
Kansas	267	5,820	4.59%
Kentucky	431	9,076	4.75%
Louisiana	104	5,136	2.02%
Maine	560	4,427	12.65%
Maryland	279	6,850	4.07%
Massachusetts	1,310	16,926	7.74%
Michigan	385	33,410	1.15%
Minnesota	1,423	16,858	8.44%
Mississippi	1,472	5,468	26.92%
Missouri	1,089	21,506	5.06%
Montana	244	2,477	9.85%
Nebraska	190	4,000	4.75
Nevada	84	5,539	1.52%
New Hampshire	349	2,609	13.38%
New Jersey	53	13,054	0.41%
New Mexico	155	2,788	5.56%
New York	280	45,400	0.62%
North Carolina	197	22,203	0.89%
North Dakota	402	3,267	12.30%
Ohio	6,814	32,180	21.17%
Oklahoma	520	7,060	7.37%
Oregon	191	4,635	4.12%
Pennsylvania	378	29,554	1.28%
Rhode Island	0	2,532	0.00%
South Carolina	286	6,500	4.40%
South Dakota	53	2,327	2.28%
Tennessee	1,430	10,524	13.59%
Texas	0	33,908	0.00%
Utah	720	4,474	16.09%
Vermont	280	1,612	17.37%
Virginia	1,358	9,193	14.77%
Washington	1,007	7,170	14.04%
West Virginia	660	3,574	18.47%
Wisconsin	772	20,678	3.73%
Wyoming	0	928	0.00%
Territories (5) , Native American Tribes (2)	0	7,891	0.00%
Total (58)	30,297	586,015	5.17%

Source: Source: Department's National Weatherization ARRA Snapshot Report dated February 16, 2010.

Department of Energy
Washington, DC 20585

February 16, 2010

MEMORANDUM FOR:	GREGORY H. FRIEDMAN INSPECTOR GENERAL
FROM:	KATHLEEN B. HOGAN (508 VERSION, NO SIGNATURE) DEPUTY ASSISTANT SECRETARY FOR ENERGY EFFICIENY OFFICE OF TECHNOLOGY DEVELOPMENT ENERGY EFFICIENCY AND RENEWABLE ENERGY
SUBJECT:	Comments on the DRAFT Special Report on "Progress in Implementing the Department of Energy's Weatherization Assistance Program under the American Recovery and Reinvestment Act"

The Department's Office of Weatherization and Intergovernmental Program (OWIP) welcomes the opportunity to review and comment on the DRAFT special report on "Progress in Implementing the Department of Energy's Weatherization Assistance Program (WAP) under the American Recovery and Reinvestment Act (ARRA)."

The Department continues to ramp up to achieve the full impact of this important program. The graphs below show the nearly exponential increase in funds spent for the national program and the number of units weatherized for selected states.

The Department's OWIP has worked diligently to achieve robust spending while maintaining proper programmatic oversight. We have worked closely with States in incorporating WAP Recovery Act requirements not previously required (e.g., Davis-Bacon Act prevailing wage provisions). After prevailing wage rates were established by the U.S. Department of Labor in September 2009, States then needed to incorporate these rates into their contracts with local agencies that perform the weatherization work. For WAP, some States had as many as 30 contracts that needed modification to incorporate revised wage rates, the requirement for weekly pay, and other changes. With States facing severe budget shortfalls, and furloughs in some cases, it was challenging for them to provide the support services (e.g., human resource departments, legal) needed to put funding mechanisms in place for Recovery Act Programs and then to spend quickly.

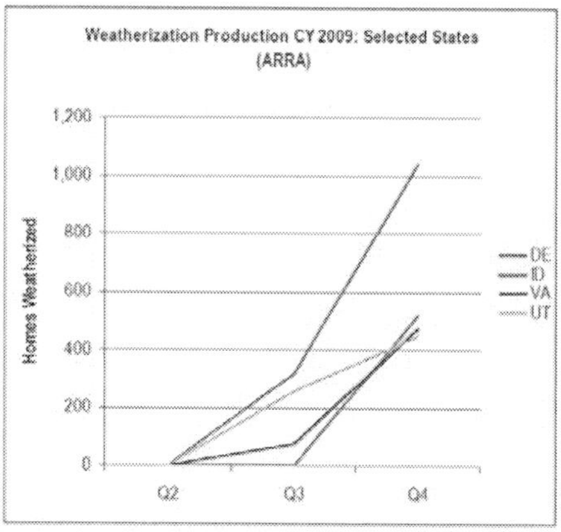

States stepped up and weatherized 36,872 homes in the fourth quarter of calendar year 2009 compared to 21,638 homes in the same quarter in 2008. The change from 2008 to 2009 represents a 70% increase in unit production. The increase in unit production tells only part of the story of improvement, given that the maximum allowable expenditure per home more than doubled from $3,200/home to $6,500/home. With the increased expenditure level, the national weatherization network performed more weatherization services on the average home in 2009, while also reaching significantly more homes than in previous years.

The Department continues to take proactive steps to accelerate the program schedule. For instance, we have developed a national agreement on Historic Preservation. The standardized template, developed for all States to use, will streamline processes for States in dealing with Historic Preservation Offices. We have also streamlined the eligibility of multi-family dwellings by adopting an established Department of Housing and Urban Development procedure. Already for Program Year 2010 (which began October 1, 2009), the Department's OWIP has issued 12 program guidance documents to provide grantees with clarification on important topics, including the implementation of the Davis-Bacon Act, the protection of historic properties, monitoring requirements, grantee performance needed before drawing down the remaining 50% of their Recovery Act WAP funding, and maintaining the privacy of recipients of services.

The Department continues to put measures in place to accelerate the program while mitigating wasteful spending risks associated with faster spending pressure. The Department's OWIP has implemented a new and improved accounts management and customer service focus, hired dozens of Project Officers (POs) to oversee State performance, and revamped its monitoring processes. The POs have formalized weekly check-ins with their grantees to enable DOE to identify and quickly respond to issues as they arise, and the POs will visit thousands of completed homes to review work products and assess local quality control efforts. Furthermore, performance-based milestones are in place. For example, States must weatherize a minimum of 30% of their total units before drawing down the remaining 50% of their Recovery Act WAP funding.

To support grantees, the Department's leadership has been actively involved in identifying and addressing challenges facing States. An outreach initiative entitled "Operation Green Light," dispatched senior leadership to visit and assess nine of the highest risk states. In addition, Assistant Secretary Cathy Zoi made telephone calls to fifty-one state energy offices—calls that in many cases included participation from state governors—to identify barriers and provide guidance to ensure that money would be spent well and quickly. Since November 19, 2009, the Department's OWIP leadership has also held weekly calls with the National Association for State Community Services Programs (NASCSP) and the National Community Action Foundation (NCAF) to further assist grantees and the weatherization assistance network in identifying and resolving issues that hinder effective and quick program spending.

In summary, the Department is confident that the weatherization efforts under the Recovery Act continue to ramp up and will achieve the economic impact envisioned while creating clean energy jobs across the nation. There is more specific feedback in the Attachment. Thank you again for the opportunity to comment.

INDEX

D

H

I

J

K

L

M

N

O

P

Q

R

U

V

W

Y